PROGRAMMING FANUC ROBOTS for Industry Applications

atp AMERICAN TECHNICAL PUBLISHERS
Orland Park, Illinois

James W. Gruenke

American Technical Publishers Editorial Staff

Editor in Chief:
Peter A. Zurlis
Director of Product Development:
Cathy A. Scruggs
Assistant Production Manager:
Nicole D. Burian
Technical Editor:
Julie M. Welch
Supervising Copy Editor:
Catherine A. Mini
Copy Editor:
James R. Hein
Editorial Assistant:
Erin E. Bello
Art Manager:
Sarah E. Kaducak

Cover Design:
Kevin L. Cuasay
Illustration/Layout:
Kevin L. Cuasay
Bethany J. Fisher
Richard O. Davis
Nicholas W. Basham
Melanie G. Doornbos
Nick G. Doornbos
Digital Media Manager:
Adam T. Schuldt
Digital Resources:
Cory S. Butler
Tim A. Miller
James V. Cashman

1 2 3 4 5 6 7 8 9 – 21 – 9 8 7 6 5 4 3

Printed in the United States of America

ISBN 978-0-8269-3412-3
eISBN 978-0-8269-9598-8

This book is printed on recycled paper.

About the Author

James (Jim) W. Gruenke

After earning his BS Industrial Technology, Plant Engineering degree from the University of Wisconsin-Stout, Jim joined the US Army. Quickly rising through the enlisted ranks, Jim was assigned as the supervisor of the Heavy Equipment and Power Generation Maintenance Shop in conjunction with the Cold Regions Test Center located in Alaska.

Leaving the military in 1989, Jim accepted the position of plant engineer and maintenance manager for a company in his home state of Wisconsin. Over the next 25 years, he held a variety of positions with four different companies and always gained responsibilities. These positions provided experience around the world as a project manager for the construction, rebuilding, and startup of facilities using innovative automation equipment to ensure high and consistent product quality.

After nearly 25 years in industry, Jim accepted an instructor position at Lakeshore Technical College (LTC) in 2009. He also returned to UW-Stout where he earned an MS in Career and Technical Education. Being tasked with advancing LTC's robotics program, Jim led a team that developed the courses and labs into a renowned program with state-of-the-art equipment and facilities. Jim currently holds the position of program coordinator and instructor in the Automation Technology and Electro-Mechanical Maintenance Technician Associate Degree programs and the Mechatronics Technician Apprenticeship program.

TECHNICAL ASSISTANCE

The author and publisher are grateful for the technical assistance provided by FANUC America Corporation.

Contents

CHAPTER 4

STARTING PROGRAMS 67

CHAPTER 5

COMPARING MOVEMENT TYPES 95

CHAPTER 6

INSTRUCTION LINES 113

Contents

Book Features

Programming FANUC® Robots for Industry Applications is an introduction to programming FANUC® industrial robots using the handheld teach pendant. This user-friendly text/workbook describes the robot cell environment, robot components, basic operation, safety considerations, and programming concepts. Step-by-step procedures walk the learner through the elements of setting up, developing, and modifying programs. These programming skills are applicable for assembly, pick-and-place, welding, and a variety of other industrial processes.

Learners will be able to establish robot frames, determine the necessary parameters for an instruction, and implement features for enhanced operations. Review questions at the end of each chapter assess comprehension. Lab exercises provide step-by-step instructions for guided practice of selected procedures. Online demonstration videos provided in the complimentary learner resources illustrate various programming commands and parameters.

Chapter introductions preview content to be covered.

Objectives identify learning goals for the chapter.

Key terms list the important vocabulary terms introduced.

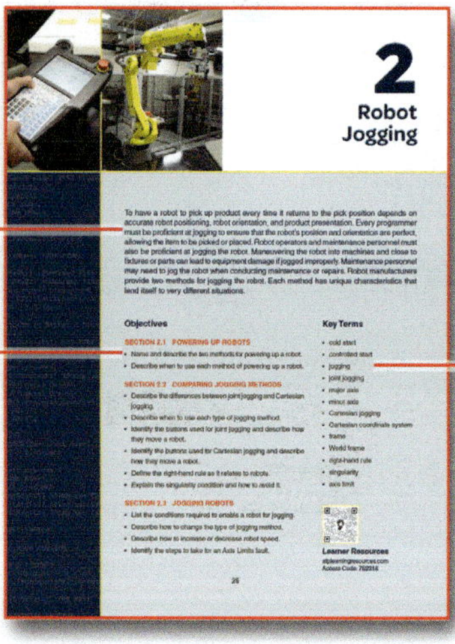

Photos depict robot and cell components.

Procedures list each step for performing common programming tasks.

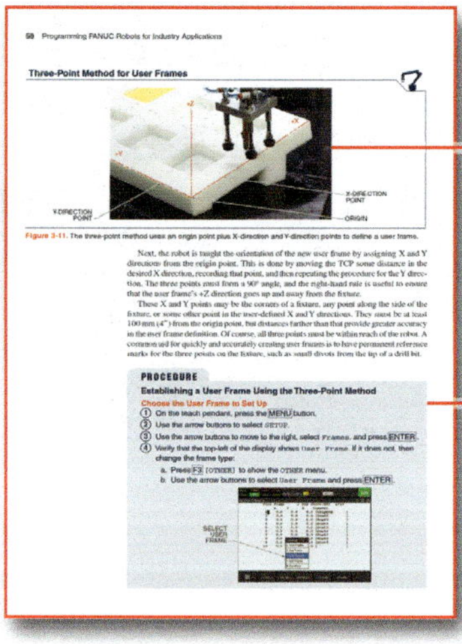

Review questions check for comprehension of chapter content.

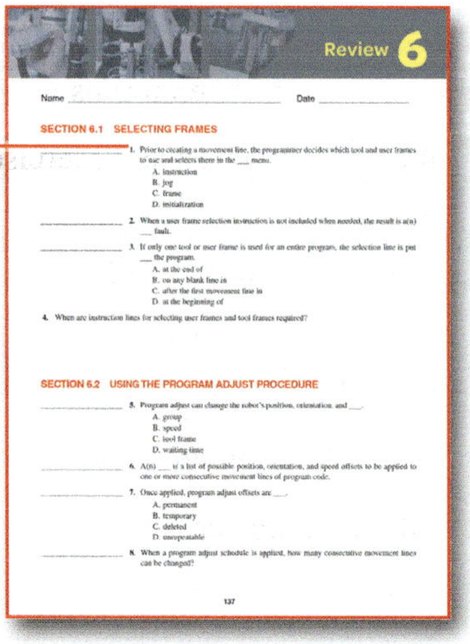

Lab exercises provide the opportunity to practice procedures on actual equipment.

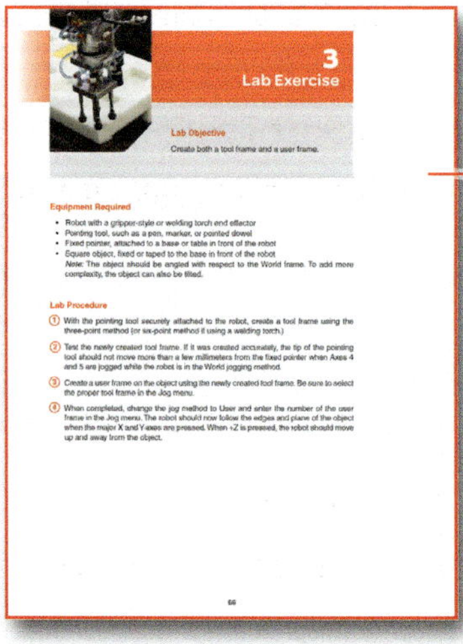

Online Learner Resources

Programming FANUC Robots for Industry Applications includes access to online learner resources that reinforce content and enhance learning. These online resources can be accessed using either of the following methods:

- Key **atplearningresources.com** into a web browser and enter Access code **762316**
- Use a Quick Response (QR) reader app to scan the QR Code with a mobile device.

The Learner Resources include the following:

- **Quick Quizzes®** that provide interactive questions for each section, with embedded links to highlighted content within the text/workbook
- **Flash Cards** that provide a self-study/review of common terms and definitions
- **Lab Exercises** that consist of printable PDF files of the end-of-chapter lab exercises and are convenient for in-lab use
- **Reference Sheets** that include useful descriptions and menu navigation hints for common programming functions
- **Media Library** that consists of videos that reinforce text/workbook content
- **Internet Resources** that provide access to additional resources to support continued learning

To obtain information on related training materials, visit **atplearning.com.**

The Publisher

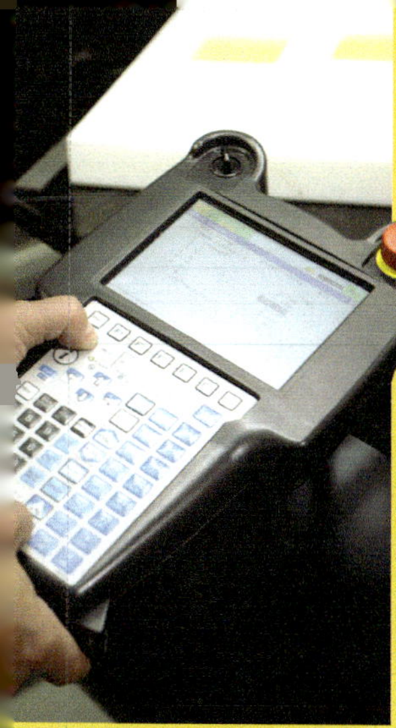

1

Robot Cells and Safety

Industrial robots have become a critical part of manufacturing and other industries because they perform repetitive tasks faster and more precisely than people can. They can even work in environments where people cannot. However, robots are complex systems that require personnel to program, operate, and maintain them for optimal performance. They also present safety hazards, which require equipment and procedures to be managed. There are relatively few regulations for the robot industry currently, but there are safety standards that are widely adopted and often enhanced by even more restrictive rules and procedures at individual facilities.

Objectives

SECTION 1.1 INDUSTRIAL ROBOTS

- List common applications for industrial robots.
- Compare the use of KAREL and teach pendant programming.
- Compare career opportunities in industrial robotics.

SECTION 1.2 ROBOT CELL COMPONENTS

- List the basic parts of a robot cell.
- Describe the role of the robot controller.
- Describe the purpose and features of the teach pendant.
- Explain how programs are created and loaded onto a robot controller.
- Describe other components or equipment that a robot may interact with inside a cell.
- List the safety features of fencing, emergency stops, and status signals.

SECTION 1.3 ROBOT SAFETY STANDARDS

- Differentiate between the requirements of OSHA and the recommendations of ANSI-RIA R15.06-2012.
- Describe the three-step process of designing a safe robotic cell.
- Describe the types of procedures required for robot cell operation and recovery.
- List other potential safety regulations that may apply to certain situations.

Key Terms

- industrial robot
- robot operator
- automation technician
- maintenance technician
- robot cell
- robot arm
- axis
- servomotor
- end effector
- tool center point (TCP)
- controller
- teach pendant
- deadman switch
- fixture
- robot reach
- emergency stop (E-stop)

Learner Resources
atplearningresources.com
Access Code: **762316**

SECTION 1.1 INDUSTRIAL ROBOTS

An *industrial robot* is an automated, programmable, mechanical actuator capable of performing tasks repeatedly with a high degree of precision. Industrial robots are an indispensable part of modern manufacturing and assembly facilities that improve production efficiency and quality. They are also highly complex systems that require highly skilled personnel for programming, operating, and maintaining them.

Robot Applications

Robot applications can include nearly any industrial task. Robots can spot weld, arc weld, assemble, paint, and load, unload, and stack parts. **See Figure 1-1.** A robot may have more than one task, such as assembly followed by loading and stacking parts. Robots accomplish these tasks with speed, precision, and endurance.

Early on, automobile manufacturing was the primary industry utilizing robots. However, in recent decades, the use of robots in manufacturing has greatly broadened in scope. Today, almost every industry uses robots in some form.

Over the years, robot technology has also changed immensely. Early robots used hydraulic actuation. Electric motors were not yet capable of both the precision positioning and force handling required for most robot applications. But as electric motor and controls technology improved, it became the primary actuating type for industrial robots.

Industrial Robots

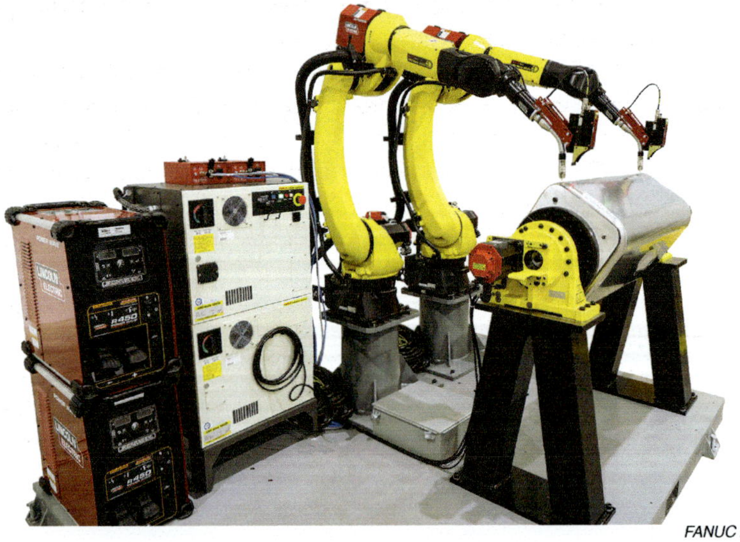

Figure 1-1. Robots are used in a variety of industries and for a variety of tasks.

Robot Programming

Robot programming has gone through many changes. All robot manufacturers have their own programming language. However, most of these languages are derived from some common base language. For example, ABB® robots use a language called Rapid, which is derived from a programming language called C. FANUC's programming language is called KAREL® (pronounced "Carol"), which is derived from the Pascal language. **See Figure 1-2.**

FANUC Robot Programming

```
$MOTYPE = JOINT
$TERMTYPE = FINE
$ORIENT_TYPE = RSWORLD
$SPEED = 1000.0
MOVE TO start_pos
direction.x = 300
direction.y = 0
direction.z = 0
MOVE RELATIVE direction
```

KAREL CODE

```
UFRAME_NUM=0
J  P[1:] 50% FINE
```

TEACH PENDANT (TP) PROGRAM

Figure 1-2. FANUC robots are programmed either in a language called KAREL or with a teach pendant.

Many years ago, a programmer needed to know programming languages, such as Rapid and KAREL, to understand how to program a robot and make it complete the desired tasks. Today, robot interfaces have become more user-friendly and often enable programmers to sequence tasks into complex programs without the need to learn the underlying programming language. For FANUC robots, this is done with the handheld teach pendant (TP) interface, and the programs are called TP programs.

However, some applications may require the type of specific control that can only be accomplished by programming directly in the robot language. One example is in automobile manufacturing, where the assembly robots need to be able to follow the continuous moving assembly line. This is a complex situation that typically requires complicated KAREL programming. Gradually though, FANUC is addressing these special applications with separate software add-ons to help develop the necessary programs with simpler programming tools. The moving assembly line situation can be programmed with an add-on module called Line Tracking. Other software add-ons are available for developing robot programs to complete other complex tasks.

Not only has robot software improved but also the software interface. Using the robot teach pendant, a programmer can program a robot to complete any possible task. The FANUC iPendant® has a large touchscreen and keypad that provides the programmer with the necessary options to program most tasks in industry.

Robot Careers

There are many different careers within the robot industry. Each career requires different levels of education and skills.

Robot Operators. A *robot operator* is a worker who readies a robot to perform programmed tasks, monitors its operation, and does any manual chores required when the robot completes a production run. Depending on the task, the operator may be responsible for placing and removing parts that the robot interacts with. Operators may be in charge of one or more robots during production. If a problem occurs, the operator works with robot or automation technicians to troubleshoot and remedy the issue. This job is an entry-level position that may lead to further education in robotics and automation.

Automation Technicians. An *automation technician* is a robot technician who programs and configures robots and may also help troubleshoot problems in robot operation. **See Figure 1-3.** An automation technician must have an in-depth knowledge of robot programming, robot capabilities, and related safety practices and requirements. They must also have a good understanding of production processes and be able to work with other production experts in order to improve product quality, production speed, and process variability. In most cases, this person has at least an associate degree in automation or a bachelor degree in some form of engineering.

Maintenance Technicians. A *maintenance technician* is a technician who maintains, repairs, and troubleshoots a robot's mechanical and electrical systems. This career requires good mechanical and electrical skills and attention to detail. Typical responsibilities include general preventive maintenance such as changing batteries, changing oil, greasing, and testing for gear backlash. In addition, this person replaces worn parts, such as gears and motors. Troubleshooting may include both mechanical and electrical issues, so a variety of skills is needed. This person should also have some degree of programming and robot-operating knowledge. This is required to test the robot after repair and place it back into service.

Automation Technicians

FANUC America

Figure 1-3. Automation technicians are involved in programming, configuring, operating, and troubleshooting industrial robots.

SECTION 1.2 ROBOT CELL COMPONENTS

All robot tasks occur within a robot cell. A *robot cell* is an all-inclusive system consisting of one or more robots and their controllers within an enclosed work area and includes any safety devices and other equipment that the robot(s) interact with. **See Figure 1-4.** A robot cell is defined by the task carried out by the robot(s) within, such as a welding cell or painting cell.

A robot cell also defines the work area of a robot. Some type of barrier, such as a fence, is typically used around the cell perimeter to prevent people from entering the cell. This is a significant safety feature because a robot is easily capable of injuring a person.

Robot Cells

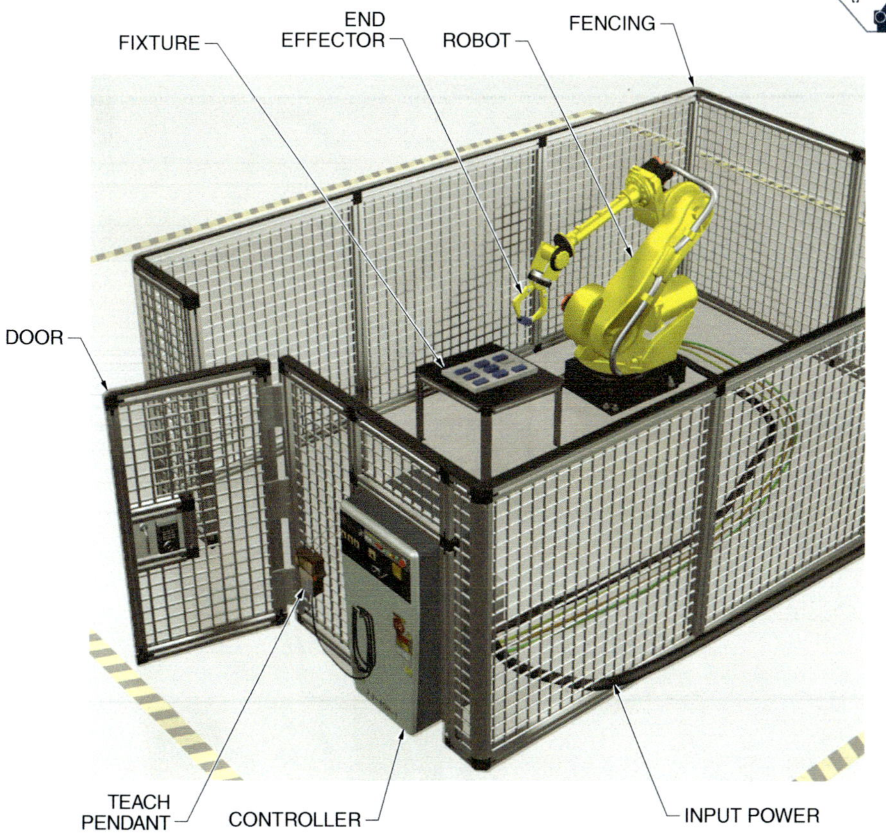

Figure 1-4. A robot cell contains all robot components and the enclosed working area.

Robots

Robots come in many different styles but can be generally categorized into a few types. **See Figure 1-5.** Each of these robot types has advantages and disadvantages.

The most common type in industry is the robot arm. A *robot arm* (or articulated robot) is a robot composed of multiple rotating joints connected by rigid segments. An *axis* is a linear or rotational direction of movement. It is common for a robot arm to have six axes (or degrees-of-freedom), but depending on the application, some may have as few as two or as many as ten. The robot arm provides strength and a long reach. It is ideal for welding and assembly tasks with larger components.

Another common type is the delta robot. A *delta robot* is a robot with a robot base and three articulated arms suspended from the base in a triangular pattern. Delta robots also consist of four or more axes. Delta robots are very fast but have a relatively short reach and are limited to lightweight applications. They are often used for pick-and-place applications for circuit board components.

The third common type is a selective compliance articulated robot arm (SCARA) robot. A *selective compliance articulated robot arm (SCARA) robot* is a robot with articulated arms that are limited to movement in two dimensions. The third dimension (up and down) may be provided by an assembly at the end of the arm that can pick and release items. This type of robot is fast and has a small footprint, being able to fold up compactly when not in use. It is often used for fine assembly work.

Robot Types

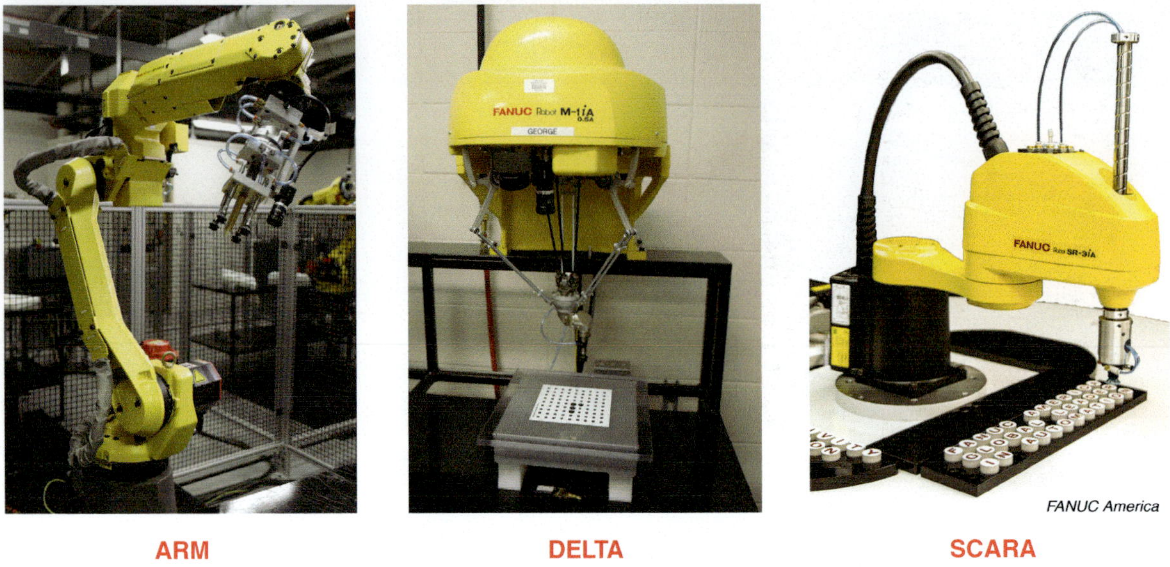

Figure 1-5. Most industrial robots can be classified into a few general types.

Each robot axis is controlled by one or more servomotors. A *servomotor* is an electric motor with an integrated position encoder that is used to control the precise degree of rotation of an axis. **See Figure 1-6.** Each axis is identified by a number. On a robot arm, the axes are numbered starting with Axis 1 at the base of the robot and progress to the end of the arm. For robot arms, the first three axes (Axes 1, 2, and 3) are called major axes, and the second three (Axes 4, 5, and 6) are called minor axes. On each axis is a witness mark. A *witness mark* is a physical mark on one side of a joint that is used with degree markings on the other side of the joint to identify the zero position of that axis. **See Figure 1-7.**

Servomotors

Figure 1-6. Each numbered axis is operated by a servomotor.

Witness Marks

Figure 1-7. Witness marks identify the zero position of each axis.

Robots are intended to be configurable for a variety of tasks. From the manufacturer, the working end of a robot arm, which is Axis 6, features a connector for attaching a variety of end effectors. **See Figure 1-8.** An *end effector* (or end-of-arm-tooling [EOAT]) is a tool attached to the end of a robot arm for completing a specific task. Common types of end effectors include grippers, pneumatic pickers, and welders.

End Effectors

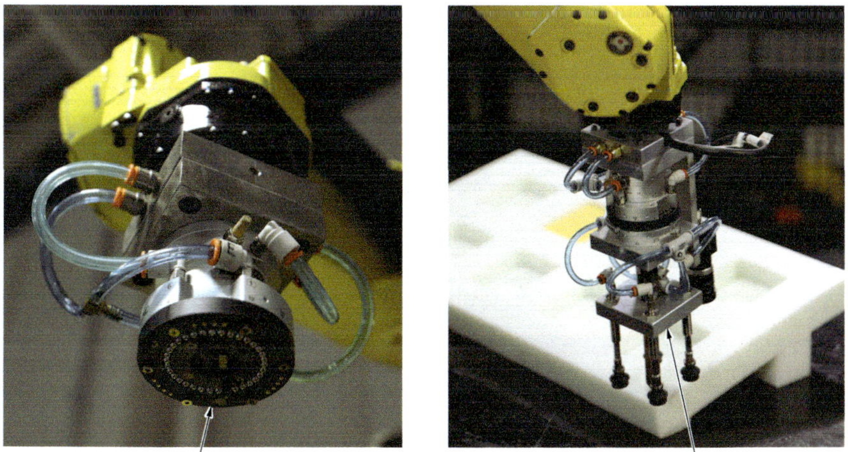

Figure 1-8. The robot arm ends with an Axis 6 connector, to which a variety of end effectors can be attached. A TCP is then defined for each specific end effector.

The robot defines its position in space in reference to its tool center point (TCP). The *tool center point (TCP)* is a point centered on an end effector and defines where the robot's work is done. From the factory, a robot has no end effector yet, so the TCP is defined as a point on the end of Axis 6. However, as soon as an end effector is attached to the robot, this point is useless, and the TCP must be redefined for the tool. Because a robot may use different types of end effectors, the TCP is redefined for each one.

Robots often move heavy items and must maintain their position accurately, so they must be securely mounted to a base. Robots are typically mounted to a stable floor foundation. However, robots can also be mounted in many other ways, such as from a ceiling, on conveyors, sideways, or at a unique angle. Many considerations go into determining the mounting type. Such things as space, reach, and obstacles are closely reviewed prior to selecting a mounting type. The robot supplier must know the orientation of the robot prior to the robot being ordered. The robot must be specifically designed for mounting sideways or from a ceiling.

Controllers

The "brain" of a robot is the controller. A *controller* (or control cabinet) is the power and control unit for a robot. **See Figure 1-9.** This is where the main robot power enters, which both runs the computer, drives, and memory inside the controller, and also feeds power and control signals to the robot servomotors. The robot computer is a type of programmable logic controller (PLC). A *programmable logic controller (PLC)* is a computer system used for controlling industrial processes.

When turning on power to the controller, it boots up much like a personal computer, completing a system check that ensures that all components are powered properly and communicating with the PLC. There are a few indicators and switches on the controller, but most of the user interface involves the teach pendant.

Each FANUC robot is identified with an F-number. The F-number on the robot must match the F-number on the controller. This ensures that the PLC and drives match the equipment on the robot arm.

Controllers

Figure 1-9. A controller is the power and control unit for a robot.

Teach Pendants

A teach pendant is connected to the controller. A *teach pendant* is a handheld device that interfaces with controller software and is used by the robot programmer to jog, configure, or program a robot. **See Figure 1-10.** The teach pendant includes a keypad and a color touchscreen display. It also includes an ON/OFF switch and an emergency stop (E-stop) button. The buttons are used to change settings, input data, and navigate through a sophisticated menu system for both programming and configuring robot operation.

Teach Pendants

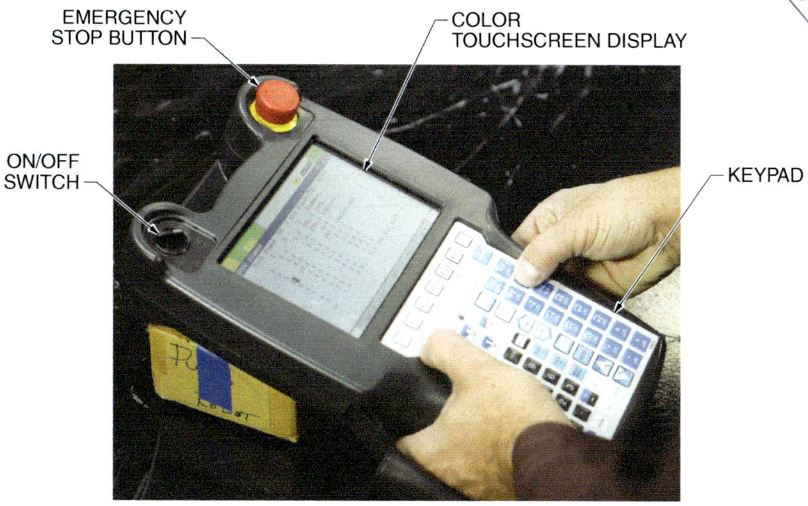

Figure 1-10. A teach pendant is used to jog, configure, or program a robot.

The teach pendant is a large device for handheld use. It includes many buttons, but there are still not enough for every programming function. The teach pendant design saves space by using the SHIFT button and function buttons.

SHIFT Buttons. The SHIFT button is used in conjunction with other buttons to modify the resulting action. Just as with a computer keyboard, SHIFT is held briefly while another button is pressed, and then both are released. Not all buttons change their function with the SHIFT button. Those that are modified are blue.

Function Buttons. The function buttons, numbered F1 through F5, are positioned directly under the display. These are buttons that change their function depending on the context (what mode or menu the teach pendant is in). The lowest part of the display contains five rectangles that may contain text assigned to the function buttons. **See Figure 1-11.** Depending on the context, some function buttons are not needed, so the text area is blank. Others contain text for functions or menus that change as needed.

It may be useful to note that functions that bring up menus (such as F2 `[METHOD]`) are in brackets, while other actions (such as F2 `DETAIL`) are not. Also, a > symbol in the lower-right corner indicates more functions are available than what are shown in the five spaces. Pressing NEXT cycles through the additional functions.

Deadman Switches. Because the teach pendant is often used to operate a robot while the operator is nearby, the pendant includes a set of deadman switches. A *deadman switch* is a switch that must be manually depressed in order for a desired operation to continue, and releasing the switch immediately stops the operation. **See Figure 1-12.** This is a safety feature to ensure that the teach pendant will control the robot only with the operator's complete attention. If the operator becomes incapacitated, distracted, or just sets the teach pendant down, the deadman switch is released, and the robot will stop all movement. There are two deadman switches, left and right, on the back of the teach pendant to suit the operator's preference. Only one needs to be used at any time. In practice, an operator will likely use one of them consistently.

Function Buttons

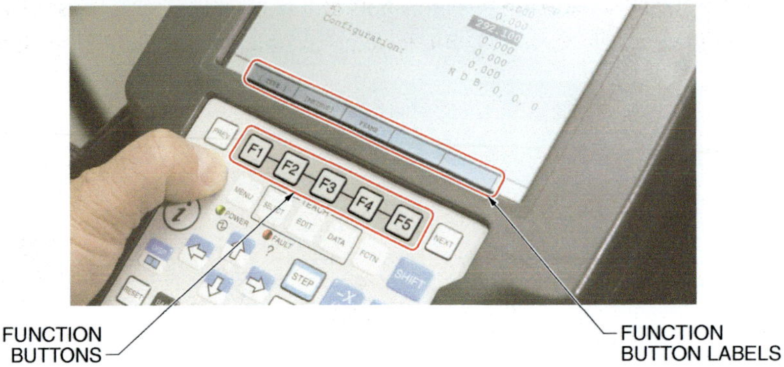

Figure 1-11. What a function button does varies depending on the mode or menu the teach pendant is in.

Deadman Switches

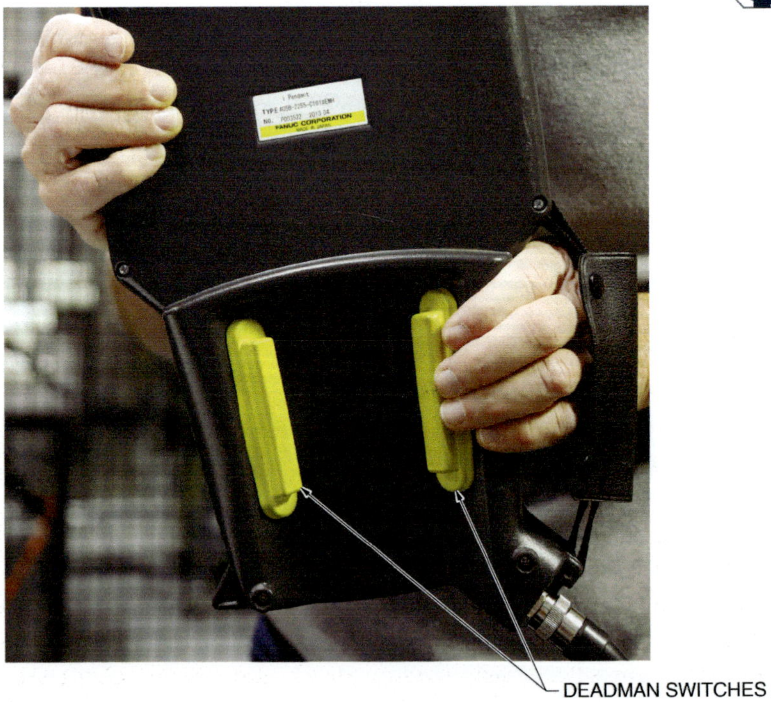

Figure 1-12. A deadman switch is a safety feature that ensures the operator's attention is on the robot.

Display Interfaces. The teach pendant display is a color touchscreen. The layout of the information on the display is generally consistent, with designated areas for fault indications, jog speed, various settings and modes, function button assignments, and the programming working area. When menus are shown, they appear on top of the programming area and may have more than one page and submenu.

Similar to personal computers, display interface designs have changed over time. Each upgrade required changes to hardware and software. The two software versions that are commonly found in current robot systems are Version 8 and Version 9, which have nearly the same capabilities but slightly different looks. **See Figure 1-13.** The examples in this book use Version 9 screenshots and instructions. Those using Version 8 will notice small differences but should have no trouble adapting the procedures.

Teach Pendant Displays

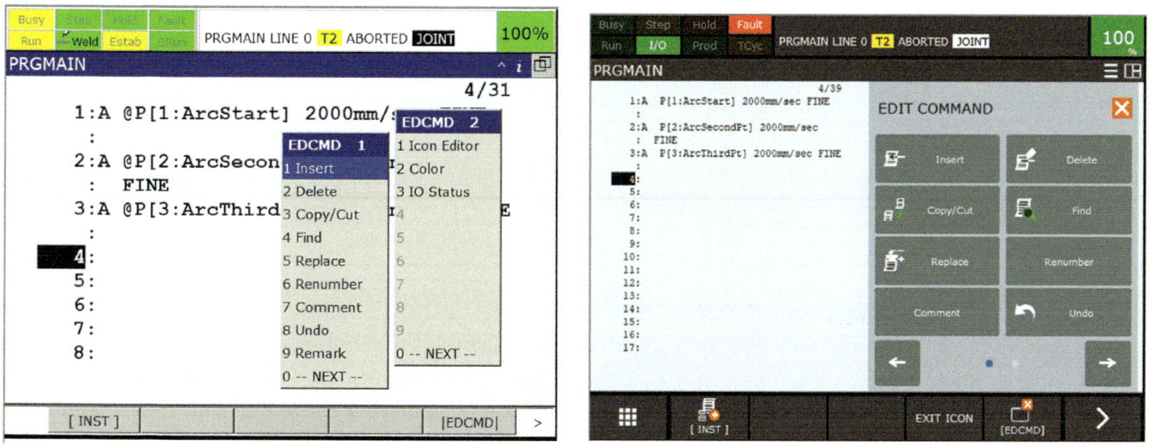

VERSION 8

VERSION 9

Figure 1-13. Version 8 and Version 9 of the teach pendant software look slightly different but have nearly the same capabilities.

Programs

Robot programs are typically created with either the teach pendant or computer software called ROBOGUIDE®. ROBOGUIDE is PC software created by FANUC that enables a programmer to lay out the entire robot cell in a virtual 3D environment. **See Figure 1-14.** Once a virtual cell is created, the programmer can then program the robot. The program files can be exported from the software onto a USB drive and then imported into the robot. The files are saved as ASCII (plain text) and binary formats.

Since programs are essentially text files, they can also be created in a text editor. However, details such as spaces and characters must be exact, or the program is unusable. Due to this complexity, programming in this way is neither typical nor recommended.

Robots can hold multiple programs. A typical robot program is comprised of a main program and several subprograms. Breaking down the robot tasks into several smaller sections in this way makes the programs easier to read and troubleshoot. As these programs are created on the teach pendant, the programs and all associated files are automatically saved.

ROBOGUIDE Software

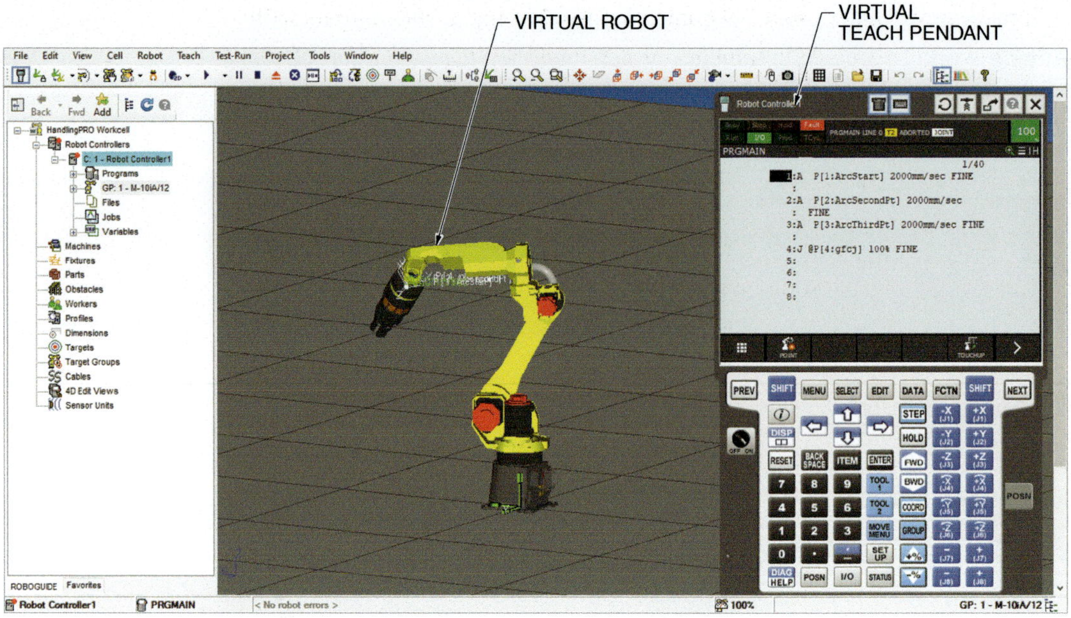

Figure 1-14. ROBOGUIDE software allows a programmer to develop programs in a virtual robot cell. The programs can then be transferred to the real robot.

Work Areas

Within a robot cell is the robot's work area. In addition to the robot and its support components, the work area may include tables and fixtures. A *fixture* is a support device for holding items or assemblies securely in a specific position. **See Figure 1-15.** By using fixtures, a robot can be programmed to move to a certain point in space to pick up a part and expect it to be there. Fixtures ensure that parts are presented to the robot in the same orientation each time. Some robots with integrated camera equipment can be programmed to find and interact with parts that are not held in fixtures, but this requires an advanced level of programming and costly hardware and software.

Fixtures

Figure 1-15. Fixtures are used to hold items in a specific position and orientation so that a robot can interact with them.

It is common for robot cells to include manufacturing equipment, such as machining centers or injection molding machines. Robots can be programmed to interact with this equipment physically. The robot controller also has inputs and outputs with which to communicate with such equipment. For example, a robot may place unfinished items into a machining center for further work. The controller signals to the machining center that the part is ready to be machined. When machining is complete, the controller receives a signal, and the robot can then safely reach into the machining center and remove the finished part.

Conveyors may be used within a work area to move items around the work area or to bring items into and out of the work area. Conveyors may move items through multiple robot cells, with each robot performing a particular task on each item.

Safety Fencing

In most cases, the robot work area is isolated with fencing. **See Figure 1-16.** The fencing is designed to prevent workers from entering a robot's work area when the robot is in operation. The fencing is also designed to prevent product from leaving the work area should it be inadvertently released from the robot end effector.

Fencing

Figure 1-16. Fencing is used both to prevent people from entering a robot's work area and to prevent items from inadvertently leaving the area.

At minimum, the fenced-in area must be large enough to enclose the robot's work envelope. The *work envelope* is the volume of space that is defined by the maximum distance the robot can reach. **See Figure 1-17.** However, other considerations may increase the size of the fenced area. Certain tasks, such as teaching, program verification, troubleshooting, and maintenance may require additional space around a robot. For an educational robot cell, there may be up to three people in the enclosure at one time (two students and one instructor), which also increases the required size of the enclosure. In industry, only one individual is allowed in a robot enclosure and only under certain conditions (the controller in manual mode and the teach pendant turned on). To allow more than one person in a robot cell, all potential energy must be locked out.

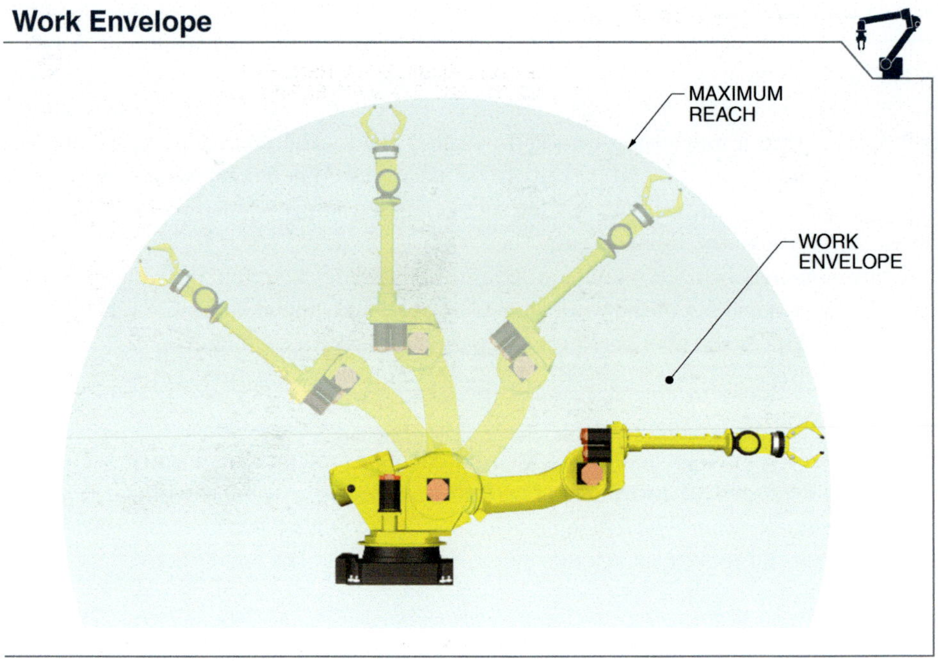

Figure 1-17. The entire space within a robot's reach is potentially hazardous for nearby personnel.

The fence material and construction are also considerations. The material may need to be modular so that sections can be removed for robot maintenance or the space enlarged for changing robot programs. Also, the enclosure must prevent unintended material from leaving the cell. For example, if a robot is grinding a part, the debris must be contained within the cell enclosure so it is not hazardous to personnel outside the cell. Another example is a robot moving boxes of parts. Should a box break open, the enclosure must not allow loose parts to escape the enclosure.

Enclosure design may also need to account for operator interaction or part inspection. If a part goes through several processes but one process cannot be automated, then the part may have to leave the cell to be processed manually and be reintroduced into the cell by the operator afterward. The cell designer must then design safe procedures for the robot and operator to transfer the part into and out of the cell.

The only time that fencing would not be required is if the robot was a collaborative robot or if the robot was contained in a room where personnel were not allowed to enter. A *collaborative robot (cobot)* is a robot that is specifically designed to operate side-by-side with personnel.

Emergency Stops

A safety fence must also prevent a person from entering the cell when the robot is in automatic mode (running its program autonomously). This is accomplished with the use of a safety door interlock switch. **See Figure 1-18.** A safety door interlock switch, when activated, causes an emergency stop (E-stop) condition. An *emergency stop (E-stop)* is a condition in which a machine's controls immediately stop all movement. A robot controller responds to an E-stop trigger by locking all brakes on the robot.

An E-stop button must also be provided for manual activation. The E-stop button has the same effect as a safety door interlock, locking all the robot brakes and immediately stopping movement within the cell.

Emergency Stops (E-Stops)

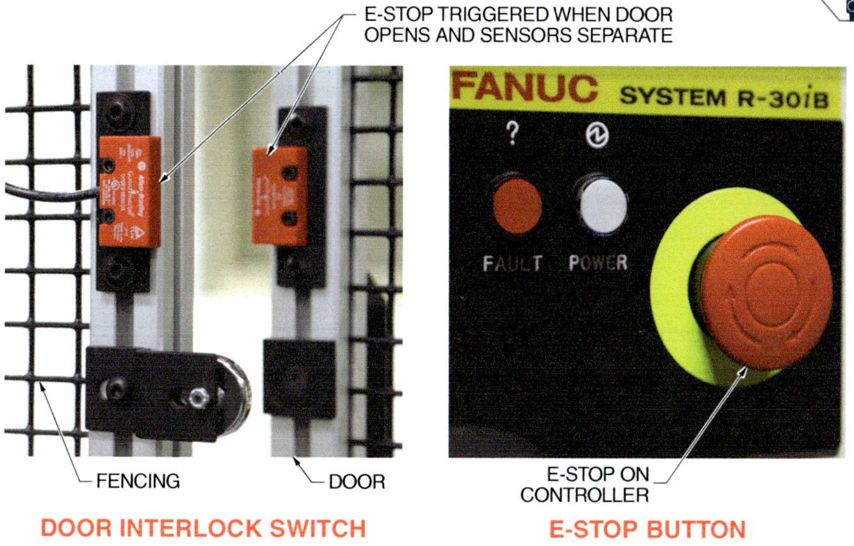

Figure 1-18. An automatic E-stop is on a fence door. A manual E-stop is also on the controller.

If there is other equipment in the cell, safety door interlocks and E-stop buttons can be wired to interrupt power to the entire cell, which stops all movement and operations within the cell. Alternatively, the safety door interlocks and E-stop buttons can be wired so that only certain pieces of equipment are interrupted. For example, in an injection molding press, the barrel of the press is filled with melted plastic that is heated using electrical power. If the power is interrupted, the plastic will solidify in the barrel and will require hours of downtime to clear. In this case, the barrel-heating circuit can likely be left out of the emergency stop action. When such partial cell emergency stops are done, a map is required with the switch to show which pieces of equipment have interrupted power and which do not.

Status Signaling

According to safety standards, visual signals must be present and visible to show the robot's condition. FANUC controllers have some indicator lights, such as a red light indicating a fault and a white light indicating that the robot's power is on.

However, it is common for facilities to develop their own visual signals and use a stacked-light system. A stack light is made up of two or more lights located one on top of the other and can be easily seen from a distance. Usually, a red light may indicate that the robot has faulted while in automatic mode. A blue light may indicate that the robot is being used in a manual mode, usually for training or for recovery from a fault. A yellow light may indicate that the robot's program is paused, such as for a part awaiting inspection. A green light may indicate that the robot is running normally in automatic mode.

SECTION 1.3 ROBOT SAFETY STANDARDS

There are few safety regulations specific to robotics. At a minimum, however, there are related OSHA regulations and voluntary safety standards developed by the robot industry. As robots are incorporated more and more into manufacturing, assembly, and other sectors, regulation may increase.

OSHA

The Occupational Safety and Health Administration (OSHA) is the governing body of industrial safety regulations. OSHA is a federal agency that has the power to enforce compliance with certain regulations and fine private-sector companies and organizations for violations. There are also many state plans working with OSHA that enforce additional regulations and/or cover additional workers excluded from the federal program.

There are currently no specific OSHA regulations for the robotics industry. However, a number of other regulations may apply, depending on the work being done by the robot. These include lockout/tagout procedures, machine guarding requirements, process-specific regulations (such as grinding, welding, or painting), and electrical safety regulations.

Robotic systems have become so sophisticated and complicated that OSHA developed a separate form (OSHA Instruction STD 01-12-002) to be completed by the OSHA inspector should they be called into a facility due to an injury or safety complaint. The form is very general, looking for negligence on the part of the employee or employer. If negligence is found, the OSHA inspector has the responsibility to demand the equipment be shut down and locked until a further inspection with a more-qualified automation engineer is completed.

ANSI-RIA R15.06

The standard ANSI-RIA R15.06, *Industrial Robots and Robot Systems—Safety Requirements,* is the primary standard for designing and operating robot cells. The development and ongoing revision process for the standard is managed by the Robotic Industries Association (RIA). This organization is composed of leading robot manufacturers, users, system integrators, component suppliers, research groups, and consulting firms. Seeing a need for robotics-specific safety requirements in the industry, they developed and continue to maintain the RIA R15.06 standard. This first version was published in 1999. The current version was released in 2012. However, many technological advances were anticipated and covered in general terms.

The American National Standards Institute (ANSI) approves of the standards-making process and designates the document as an American National Standard. This designation by itself does not compel compliance by any company or organization, but it does provide validation of the content and its relevance to the industry.

Even though compliance with this standard is voluntary from a governmental point of view, it has become very important to industry organizations. Its requirements are commonly integrated into their internal policies and procedures.

The primary content of ANSI-RIA R15.06 describes how to implement a three-step process for designing a robot cell. **See Figure 1-19.** The first step identifies the range of robot tasks and requirements. The second step recognizes the risks connected with every task performed by or on the robot. The third step is the actual process of designing the robot cell while taking into consideration all the risks identified in the second step.

Included in this chapter is the three-step process used to design a robotics lab at an educational facility. This is intended to assist in establishing a safe educational cell and provide the student with a glimpse of the attention needed to design and construct a safe robotic cell in the industry. By no means is this information intended to provide enough instruction necessary to design and construct an entire robotic cell. Much more education and experience are needed to undertake this task.

Step One: Required Robot Tasks. The first step is to identify the range of robot tasks and requirements. For an educational robot cell, the tasks may include various inspections of the robot before operation, simple maintenance and repair tasks, and programming and operation tasks. **See Figure 1-20.**

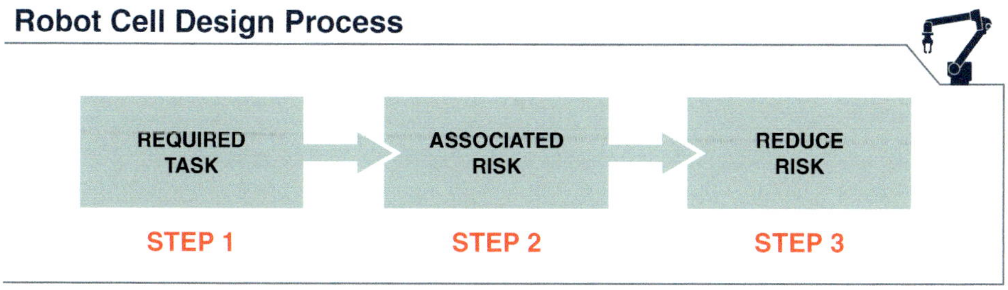

Figure 1-19. The primary purpose of ANSI-RIA R15.06 is to explain how to design a robot cell based on a risk assessment of the tasks involved.

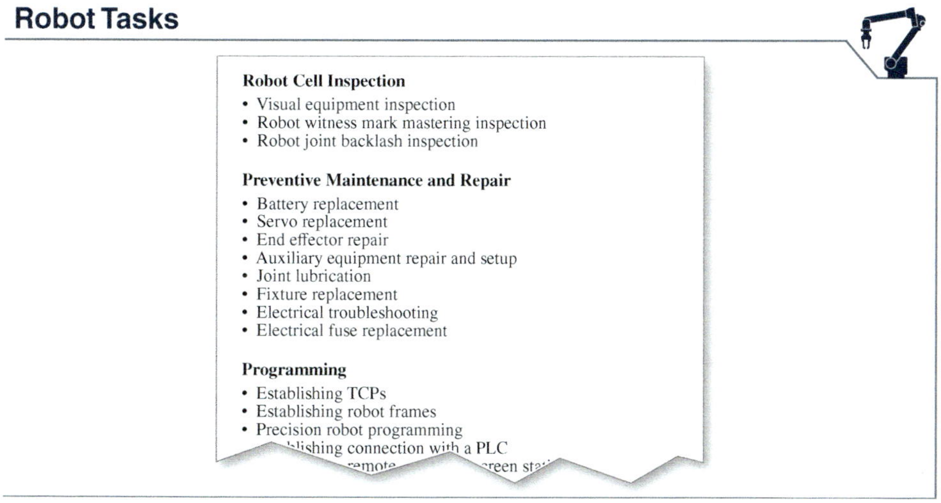

Robot Cell Inspection

- Visual equipment inspection
- Robot witness mark mastering inspection
- Robot joint backlash inspection

Preventive Maintenance and Repair

- Battery replacement
- Servo replacement
- End effector repair
- Auxiliary equipment repair and setup
- Joint lubrication
- Fixture replacement
- Electrical troubleshooting
- Electrical fuse replacement

Programming

- Establishing TCPs
- Establishing robot frames
- Precision robot programming

Figure 1-20. In step one, all tasks associated with the robot are listed.

Changing Robot Batteries

FANUC robots store certain data related to joint position in a type of memory that requires power at all times. When the robot is plugged in and the controller is turned on, power is supplied to this memory. When the controller is turned off, power is maintained to this memory by batteries in the robot's base. If the batteries fail, this joint position data is then lost when the robot and controller are turned off. When the robot is turned on again, this critical data must be reestablished through the remastering procedure.

In an industry setting, robots are seldom, if ever, turned off. For this reason, the backup batteries are often forgotten and not replaced every six months as suggested by the robot manufacturer. When the batteries are getting low, a battery warning fault is displayed during startup. The robot operator must press the RESET button to continue operations. However, once reset, the fault will not be displayed again.

Changing Batteries

1. Power up the robot.
2. Turn the key switch on the controller to the T1 position.
3. Turn the teach pendant on.
4. Carry the teach pendant and four new C-sized batteries into the robot cell to the back of the robot.
5. Remove battery cover and old batteries.
6. Replace with new batteries, ensuring the batteries are in the correct direction as shown on the battery cover.
7. Replace battery cover.

Step Two: Risks Associated with Each Task. The second step is a risk assessment. Each task is evaluated individually. The robot status is indicated in italics, and the risks associated with the tasks are listed in red. **See Figure 1-21.** All programming tasks are seen to have the same risks. This being the case, there is no reason to list the same items multiple times. The robot cell is assumed to include fencing and appropriate safety gate systems, which will limit the overall risk to others in the area. However, working within the robot enclosure are two students and an instructor. This fact must be addressed in the risk of each task.

Robot Task Risks

Robot Cell Inspection

- Visual equipment inspection

Robot in rest position; no part in end effector; electrical power locked out; air pressure on

  - Minimal risk of tripping on cable
  - Minimal risk of compressed air

- Robot witness mark mastering inspection

Robot in rest position; no part in end effector; electrical power on; air pressure on; teach pendant (TP) in cell with up to two students

  - Minimal risk of tripping on cable
  - Minimal risk of compressed air
  - Low risk of pinch between joints
  - Minimal risk of student struck by robot
  - Minimal risk of student being trapped between robot and fence

- Robot joint backlash inspection

Robot outstretched; no part in end effector; electrical power on; air pressure on; TP in cell with up to two students

  - Minimal risk of tripping on cable
  - Minimal risk of compressed air
  - Low risk of pinch between joints
  - Minimal risk of student struck by robot
  - Minimal risk of student being trapped between robot and fence
  - ...risk of bumping head on low overhang of robot arm

Figure 1-21. For step two, each task's potential risks are identified.

Step Three: Reducing Risks and Revising Safety Training. Once the risk assessment is complete, the risks are reviewed to determine if they can be reduced by substitution or safeguarding. The priorities of ANSI-RIA R15.06 are the following:

1. Elimination of the hazard
2. Safeguarding to prevent the operator from coming in contact with the hazard or to bring the hazard to a safe state prior to the operator coming in contact with the hazard
3. Supplementary measures such as warning signs, training, and personal protective equipment (PPE)

As can be seen from the risks, PPE will need to be worn during some tasks along with the completion of additional safety training to minimize specific risks. Students must be instructed to keep the table between themselves and the robot when the power is not locked out, and students must also be made aware of the robot's pinch points. Each class period, prior to releasing the students to work with the robots, all training elements involving the reduction of safety risks should be reviewed and documented.

If any task has a risk determined to be high, additional safeguards may need to be added. An example of a task in need of additional safeguards is when a robot trims a plastic part with a high-speed router. The operator must unload and reload the cell when the part is finished. However, even though the router power is off during this time, the cutter still spins at a high rate of speed for a few seconds. The integrator has several options to reduce the risk with additional safeguards:

- Prevent the door from opening until the cutter has stopped.
- Put a drive on the router motor to stop the cutter quickly.
- Put a brake on the router motor to stop the cutter quickly.
- Put the cutter in a small enclosure while the part is loaded and unloaded.

The first three options may be undesirable because they extend the process time too much, are too costly, or add too much weight to the robot. Those drawbacks do not arise with the fourth option, which will minimize the risk for the operator while loading and unloading the robot. The fourth option will not, however, reduce the risk for maintenance personnel. They will need to be trained to verify that the cutter has stopped, and a warning sign must be placed on the equipment.

Other items that should be reviewed during the initial training period are acts that are normally thought to be common sense. For example, a person should never be locked in a robot cell while the cell or robot is running automatically. It is a reminder to students that individuals sometimes act without thinking and that robots and other automated equipment can be dangerous.

Industrial Hazards. In an educational setting, the robot cells are quite simple compared to those in industry. Additional potential hazards must be considered when working with robots in industrial settings. Commonly encountered hazards include the following:

- mechanical hazards, such as movement of the robotic arm, movement of additional axes, movement of the end-effector, movement of rotational blades, end-effector separation, product ejection, inability to exit the robot cell, unintended movement, and unexpected release of potential energy
- electrical hazards
- thermal hazards
- noise hazards
- vibration hazards
- radiation hazards
- material and substance hazards
- ergonomic hazards
- poorly designed teach pendant, HMI touch screen, or operator panel (such as a panel that is too high)
- poorly designed loading/unloading arrangements
- inappropriate location of components for troubleshooting, repair, or adjustment
- hazards associated with general facility environment
- combinations of hazards

Operation and Recovery Requirements. ANSI-RIA R15.06 also requires that procedures for robot cell operation and recovery be developed and that personnel be trained in them. Each procedure should describe the beginning status of the cell and how to place the cell into the desired condition. The procedures are generally broken into several categories:

- cell cleaning
- cell setup/startup
- cell shutdown/runout
- cell fault recovery

For example, for cell cleaning, either all production parts need to be transferred out of the cell before cleaning begins, or the cell can be shut down once the last production part has entered the cell. By shutting the cell down when the last production part enters the cell, the cell is primed for startup. The procedure should also list any special tools needed and areas where special care is to be taken during cleaning.

As for setup/startup, the procedure should describe how to select the proper robot or equipment programs and how to prime the cell if the cell has run out of parts. If tool or end effector changes are required, the procedures may provide an illustration of the proper tool or end effector and describe, in detail, how to make the change.

For cell shutdown/runout, the procedures should include how the cell should be left after the robot has finished a production run or completed a shift. In some cases, certain pieces of equipment must stay powered up while others may need to be locked out.

The last type of procedure is a fault recovery. Most times, this is the most complicated procedure, requiring the operator to manually jog the robot and decide as to the condition of the cell equipment and the parts in question.

Even for cells in an educational setting, startup, programming, and shutdown procedures should exist. This ensures that the robot cell is left in the same condition as when the student arrived.

Other Relevant Safety Regulations

In addition to applicable OSHA regulations and voluntary robotic safety standards, a number of other regulations and standards may be applicable in some jurisdictions or to a particular robot cell and its operation or maintenance. It is important to consider the electrical standards NFPA 70®, *National Electrical Code*®, and NFPA 70E®, *Standard for Electrical Safety in the Workplace*. There are also a number of relevant International Organization for Standardization (ISO®) standards, such as ISO 10218-1, *Robots and robotic devices—Safety requirements for industrial robots—Part 1: Robots.*

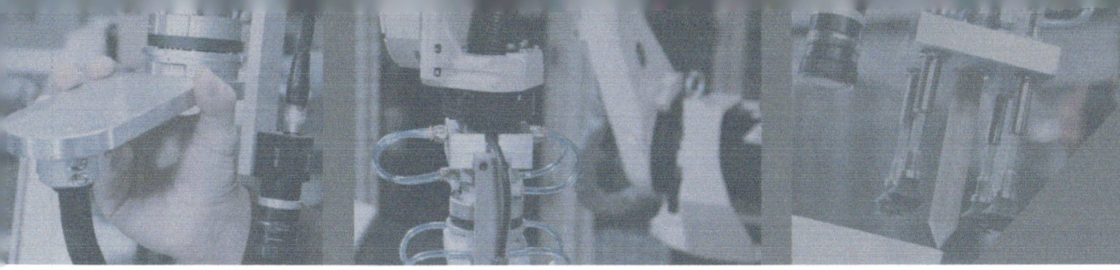

Review 1

Name ______________________________ Date ______________

SECTION 1.1 INDUSTRIAL ROBOTS

__________ **1.** With today's technology, ___ are the primary actuating type for industrial robots.

__________ **2.** FANUC's programing language is called ___.

A. C
B. Rapid
C. KAREL
D. Macro

__________ **3.** A(n) ___ technician is a robot technician who programs and configures robots and may also help troubleshoot problems in robot operation.

A. operating
B. automation
C. maintenance
D. production

4. How has programming gotten simpler for complex situations?

5. Why must a robot maintenance technician have a variety of skills?

SECTION 1.2 ROBOT CELL COMPONENTS

__________ **6.** A(n) ___ is an all-inclusive system consisting of one or more robots and their controllers within an enclosed work area and includes any safety devices and other equipment that the robot(s) interact with.

__________ **7.** A ___ is a robot composed of multiple rotating joints connected by rigid segments.

A. SCARA robot
B. servorobot
C. delta robot
D. robot arm

_______________ **8.** For robot arms, the first three axes (Axes 1, 2, and 3) are called ___ axes.

_______________ **9.** From the factory, the TCP is defined as a point ___.

A. on the end of Axis 6
B. at the intersection of Axes 1 and 2
C. at the witness mark
D. on the included end effector

_______________ **10.** A(n) ___ is a handheld device that interfaces with controller software and is used by the robot programmer to jog or program a robot.

_______________ **11.** Robot programs are typically created with either a teach pendant or ___.

A. text editors
B. PC software
C. motion capture
D. a controller

_______________ **12.** A(n) ___ is a support device for holding items or assemblies securely in a specific position.

_______________ **13.** The only time that safety fencing would not be required is if the robot was a(n) ___ robot or if the robot was contained in a room where personnel were not allowed to enter.

14. How many axes could a robot arm have?

15. What are some common types of end effectors?

16. How is power routed to a robot?

17. How do deadman switches function as safety devices?

18. Other than robot reach, what are some additional considerations for defining a robot's work area?

19. How does a robot controller respond to an E-stop trigger?

SECTION 1.3 ROBOT SAFETY STANDARDS

____________________ **20.** The standard ANSI-RIA ___, *Industrial Robots and Robot Systems—Safety Requirements,* is the primary standard for designing and operating robot cells.

____________________ **21.** The second step in designing a robot cell is ___.

A. task identification
B. hazard elimination
C. risk assessment
D. safeguarding

____________________ **22.** If any task has a risk determined to be high, additional ___ may need to be added.

____________________ **23.** For cell ___, the procedures should include how the cell should be left after the robot has finished a production run or completed a shift.

A. cleaning
B. setup/startup
C. shutdown/runout
D. fault recovery

24. Are there OSHA regulations specifically for robots?

25. What OSHA regulations may apply, depending on the work being done by the robot?

26. Is the RIA R15.06 standard required?

1 Lab Exercise

Lab Objective

Complete a safety risk assessment of selected robot tasks.

Equipment Required

None

Lab Procedure

1. Select two tasks from the Step One list shown in Figure 1-20.
2. Complete a risk assessment for each task. Include the condition of the robot and suggestions to mitigate all risks associated with the tasks.

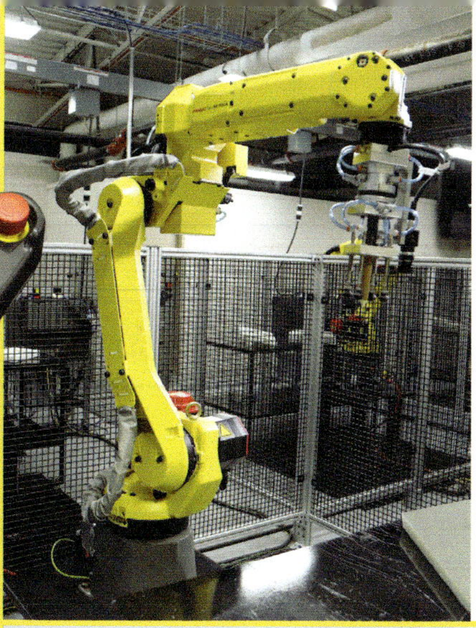

2 Robot Jogging

To have a robot to pick up product every time it returns to the pick position depends on accurate robot positioning, robot orientation, and product presentation. Every programmer must be proficient at jogging to ensure that the robot's position and orientation are perfect, allowing the item to be picked or placed. Robot operators and maintenance personnel must also be proficient at jogging the robot. Maneuvering the robot into machines and close to fixtures or parts can lead to equipment damage if jogged improperly. Maintenance personnel may need to jog the robot when conducting maintenance or repairs. Robot manufacturers provide two methods for jogging the robot. Each method has unique characteristics that lend itself to very different situations.

Objectives

SECTION 2.1 POWERING UP ROBOTS

- Describe the two methods for powering up a robot.

SECTION 2.2 COMPARING JOGGING METHODS

- Describe joint jogging and its advantages.
- Describe Cartesian jogging and its advantages.
- Describe how the Cartesian coordinate system is used to make frames.
- Define the right-hand rule as it relates to robots.
- Explain the singularity condition and how to avoid it.

SECTION 2.3 JOGGING ROBOTS

- List the conditions required to enable a robot for jogging.
- Explain how to select the method of jogging a robot.
- Describe how to increase or decrease robot speed.
- Identify the steps to take for an Axis Limits fault.

Key Terms

- cold start
- controlled start
- jogging
- joint jogging
- major axis
- minor axis
- Cartesian jogging
- Cartesian coordinate system
- frame
- World frame
- right-hand rule
- singularity
- axis limit

Learner Resources
atplearningresources.com
Access Code: **762316**

SECTION 2.1 POWERING UP ROBOTS

There are two ways to power up a robot: cold start and controlled start. *Cold start* is the standard method of initializing a robot for typical programming or operations by turning on the power while not pressing any teach pendant buttons. Cold start is done by simply turning the power switch on the controller clockwise. **See Figure 2-1.** Like a personal computer, the robot takes a minute or so to run through its system tests and then boot up the teach pendant. Once the controller is turned on, no buttons should be pressed on the teach pendant until it is completely booted. Then, the robot operator can begin to jog the robot.

Controller Startup

Figure 2-1. The power switch on the controller front panel is used to cold start a robot.

Controlled start is the startup procedure for powering up a robot in preparation for adding or modifying robot software and overriding errors. This procedure requires the operator to also hold down the PREV button and NEXT button on the teach pendant while the robot is booting. **See Figure 2-2.** When the robot controller is powered in the controlled-start mode, the controller booting process is changed, providing access to options that are not present in a standard cold-start mode. This method for booting a robot is seldom required and should not be used for everyday use.

Jogging also requires that the operating mode key switch be in the T1 position and the teach pendant ON/OFF switch to be in the ON position. This is a safety precaution. When this switch is in the ON position, the robot cannot be activated automatically, regardless of the position of the controller's key switch.

SECTION 2.2 COMPARING JOGGING METHODS

Jogging is the manual adjustment of a robot's position and orientation via the teach pendant. Utilizing the different methods of jogging and adjusting the speed of jogging, the programmer can maneuver a robot into machines and precisely orient the robot in order to place or pick objects or production parts. Jogging can be done very slowly in order to position the robot precisely.

Teach Pendant Startup

Figure 2-2. At the upper left of the teach pendant is the ON/OFF switch. This switch must be in the ON position in order to jog the robot or make changes to a program.

Joint Jogging

Joint jogging is a jogging method in which each robot joint is operated independently (with the exception of Axis 2). **See Figure 2-3.** This is also called "jogging in joint" or simply "joint." For example, Axis 1, 3, 4, 5, or 6 can be rotated in one direction by pressing the +J button on the teach pendant for that axis and rotated in the other direction by pressing the corresponding –J button.

The exception is Axis 2, which is interactive with Axis 3. This means as the operator jogs Axis 2, Axis 3 also moves. This interaction prevents the tool tip from being accidentally driven into a table or fixture when jogging Axis 2. While such crashes could potentially occur from any axis movement, the orientation of Axis 2 makes it a little more likely. **See Figure 2-4.** However, when jogging Axis 3 in joint, only Axis 3 moves.

A robot's axes are separated into two categories. A *major axis* is one of the first three axes (1, 2, or 3), which provide the largest amount of robot movement. A *minor axis* is one of the last three axes (4, 5, or 6), which provide fine robot movement. Distinguishing between major and minor axes is important when discussing movement.

Robot Axes

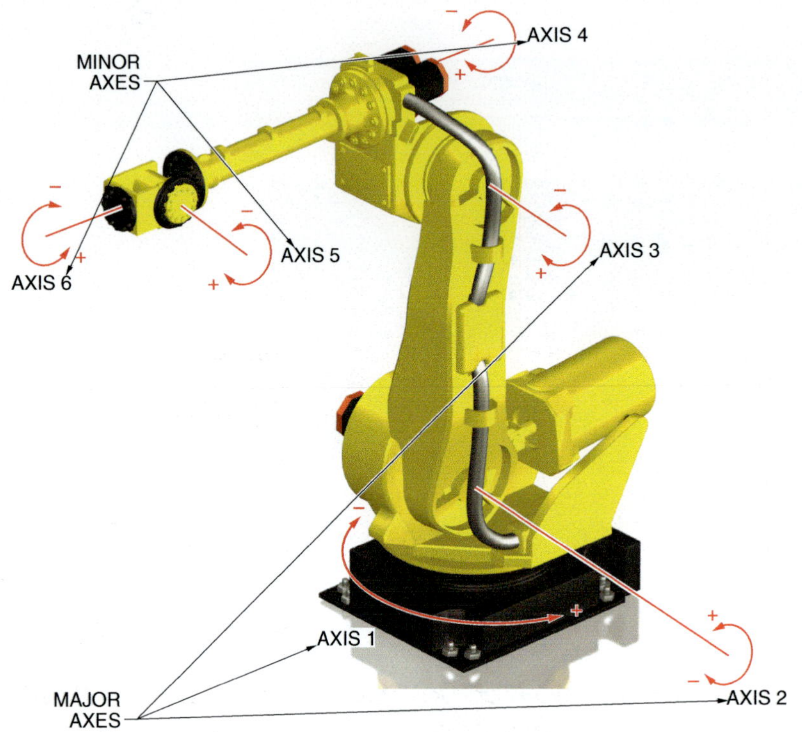

Figure 2-3. When joint jogging, each robot joint (except Axis 2) is moved independently.

Axis 2 and Axis 3 Interaction

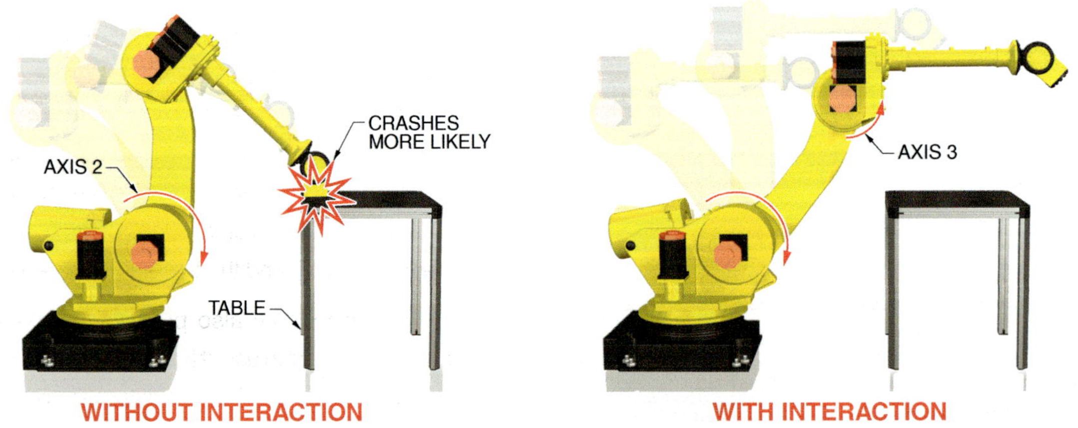

Figure 2-4. Axis 2 is interactive with Axis 3. They move together in order to reduce the likelihood of crashes.

The teach pendant actually has eight pairs of jog buttons. **See Figure 2-5.** The buttons for Axes 1 through 6 are labeled J1 through J6. Each joint has two buttons, one for each rotation direction. The last two pairs, labeled J7 and J8, are assigned to the axes of additional servomotors that may be used to move other parts of the robot or equipment the robot interacts with. These jog buttons could possibly be used for end effector components or equipment adjacent to the robot. Typically though, the J7 and J8 buttons are not used.

Joint Jogging Buttons

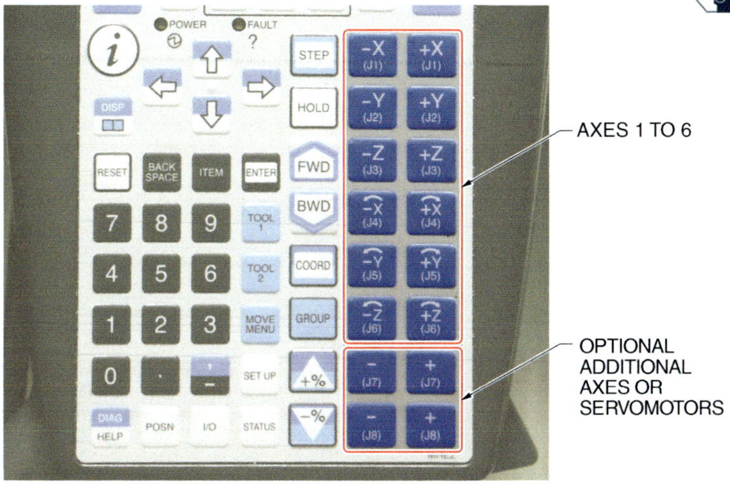

Figure 2-5. The jog buttons on the teach pendant are divided by axis and direction.

Robot Lubrication Maintenance

FANUC's maintenance manual includes procedures for lubrication and recommendations for frequency. Generally, robots should be lubricated once a year. Each robot axis requires either a grease or oil change. Some robots have one axis that contains oil, while the other axes use grease as the lubricant. (Larger robots may only have greased axes.)

The procedure for greasing is to remove two grease plugs. One port allows air to enter, while the other is used to add new grease. When new grease is added, the old grease is pushed out the air port. The old grease is typically visibly dirty. New grease is added until the color of the grease exiting the robot matches the color of the new grease. The plugs are then replaced in the ports.

A robot's oil levels must be checked periodically, which is done with a sight glass. For most industry applications, this is a weekly or monthly inspection. The oil levels should be maintained at their normal level. If the oil becomes oxidized by heat or age, the sight glass will be discolored. Should this happen, the oil must be drained and the sight glass cleaned. To check the oil level, the robot is placed in the zero position (all witness marks aligned).

Changing oil is done in a manner similar to greasing. The robot is positioned so the oil for an axis will drain. The two oil plugs for the selected axis are removed, and the oil is allowed to completely drain. Then new oil is added until it just begins to flow out the drain plug, and the bottom plug is reinstalled. Then oil is added until the axis is filled to the proper level, and the top plug is reinstalled.

The maintenance manual also provides details for the proper grease and oil to use. It is important to use the proper grease and oil in order to protect the different metals from the corrosion potential, forces, and heat that generally occur during industrial use.

Joint jogging is generally used for large movements. For example, a programmer would typically use joint jogging to move a robot from one table to another. **See Figure 2-6.** In this case, a joint move of Axis 1 would be the most efficient way to do so. While standing behind the robot, pressing –J1 jogs the robot to the right, and pressing +J1 jogs it to the left. All other axes remain in their current position.

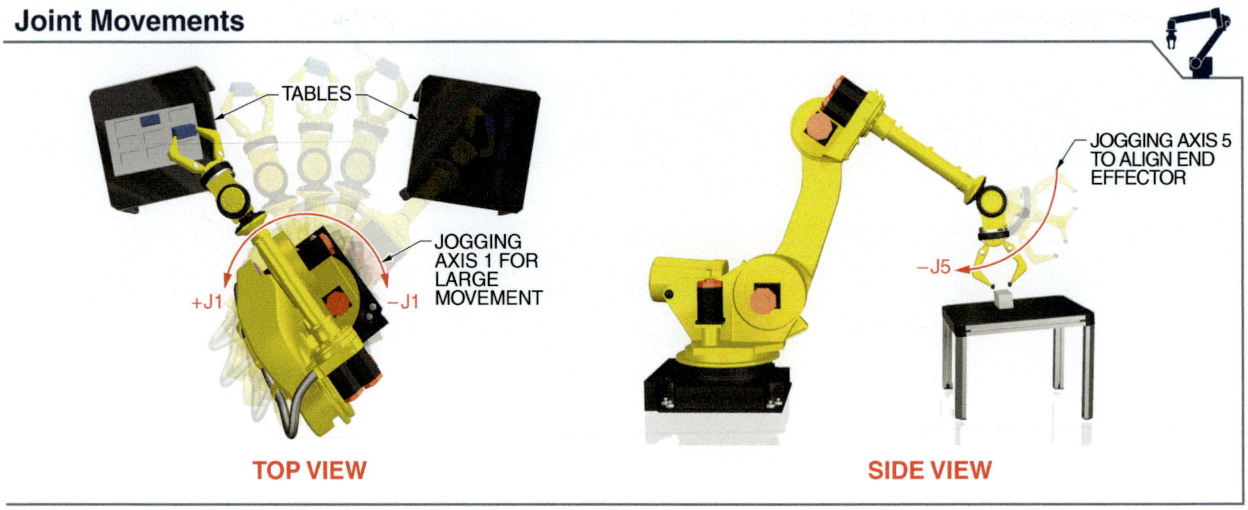

Figure 2-6. Joint jogging is particularly useful for large movements and also for adjusting the orientation of the end effector.

Joint jogging is also convenient for adjusting the orientation of an end effector to a fixture or part. For example, the end effector may be at an angle in relation to an item when it needs to be in a straight-on orientation. In this case, the operator needs to align the end effector to the table. Jogging Axis 5 will align the end effector.

Cartesian Jogging

Cartesian jogging is a jogging method in which all axes operate simultaneously (as needed) in order to move the tool center point (TCP) in a straight line or to move the robot about the TCP. There are several reasons to use Cartesian jogging. Moving in straight lines or around a point can be easier for the operator or programmer to understand and manage, especially when making tiny adjustments. Robot operators and programmers commonly use this method of jogging to orientate the robot to an object they wish to pick or to a location to place an item. The robot can also orientate a part being held in the robot's end effector to another object.

Cartesian Frames. Cartesian jogging is based on the Cartesian coordinate system. The *Cartesian coordinate system* is a system of describing points in three-dimensional space as distances along three perpendicular axes from an origin point. **See Figure 2-7.** For example, a point may be four units in one direction, one unit in a second (perpendicular) direction, and two units in a third (perpendicular to both first and second) direction. These coordinates may be written as (4, 1, 2). This system is actually quite familiar to most people, as it is similar to the latitudes and longitudes used to define locations on Earth.

For robots, the Cartesian coordinate system is used to make a frame. A *frame* is a set of three perpendicular axes (X, Y, and Z), which intersect at an origin (zero) point. Jogging the robot in reference to a frame is therefore known as Cartesian jogging.

Cartesian Movements. Using the major axis (Axis 1, Axis 2, or Axis 3) buttons, the operator can move the TCP from one location to another while it maintains its orientation. This type of movement cannot be accomplished using the joint method of jogging. The first set of jog buttons controls movement in the X axis. When in the default (World) jogging method, the +X button jogs the robot TCP in a straight line away from the robot, while the –X button moves the TCP toward the robot. The second set of jog buttons is for the Y axis and moves the TCP in a straight line perpendicular to the X axis. The third set of jog buttons is for the Z axis. The Z buttons jog the TCP up (+Z) and down (–Z).

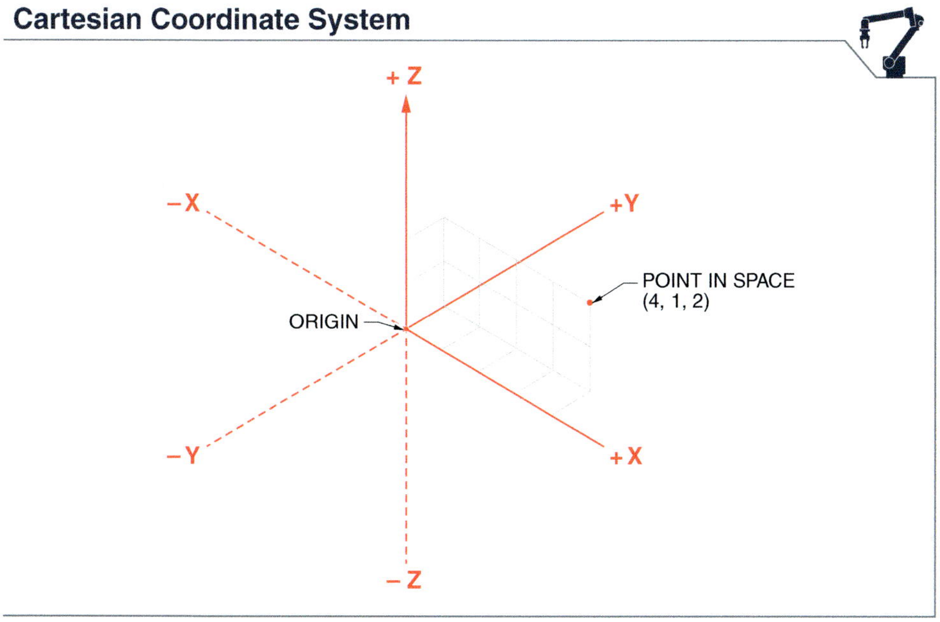

Figure 2-7. The Cartesian coordinate system is defined by three perpendicular axes, which intersect at an origin point.

Jogging the minor axis (Axis 4, Axis 5, or Axis 6) buttons while in the Cartesian jogging method produces a different type of movement. In this case, the TCP stays in place while the robot maneuvers around that point, in a direction around a Cartesian axis. If the $\overleftarrow{X}$ or $\overrightarrow{X}$ buttons are pressed, the robot rotates left or right around the X axis. The $\overleftarrow{Y}$ and $\overrightarrow{Y}$ set of jog buttons rotates the robot over the TCP, toward the front or back of the robot, around the Y axis. The $\overleftarrow{Z}$ and $\overrightarrow{Z}$ buttons rotate the robot around the Z axis.

Frame Types. There are four types of Cartesian frames to choose from when jogging a robot: World, tool, user, and jog. The origin and orientation for each of these frames can be different. The only frame preset by the manufacturer is World. The *World frame* is the default Cartesian frame in which the origin is centered on and just above the rotation point of Axis 1, the X axis is pointing out from the robot's front, the Y axis is pointing to the robot's left, and the Z axis is pointing up. **See Figure 2-8.**

Tool, user, and jog frames must be created by the programmer. The applications of these other frames and how to set them up is covered in later chapters. If the tool, user, or jog frame jogging methods are inadvertently selected before they are defined, the robot will automatically default to the World frame.

Right-Hand Rule. An easy way to remember the direction a robot will move when certain jog buttons are pressed is to use the right-hand rule. The *right-hand rule* is a hand arrangement that uses fingers as a memory aid to show the directions of the X, Y, and Z axes.

When standing behind and facing the robot, the operator points the index finger of their right hand away from their body. **See Figure 2-9.** This is the +X plane direction. The second finger is unfolded to the left, 90° to the index finger. This indicates the robot's +Y plane direction. Unfolding the thumb and pointing it up is the robot's +Z plane direction. When the index finger is pointing directly away from the front of the robot and both the index finger and second finger are parallel with the base of the robot, this is the representation of the World frame directions. The right-hand rule can also be used at other origins and orientations to represent other frames.

World Frame

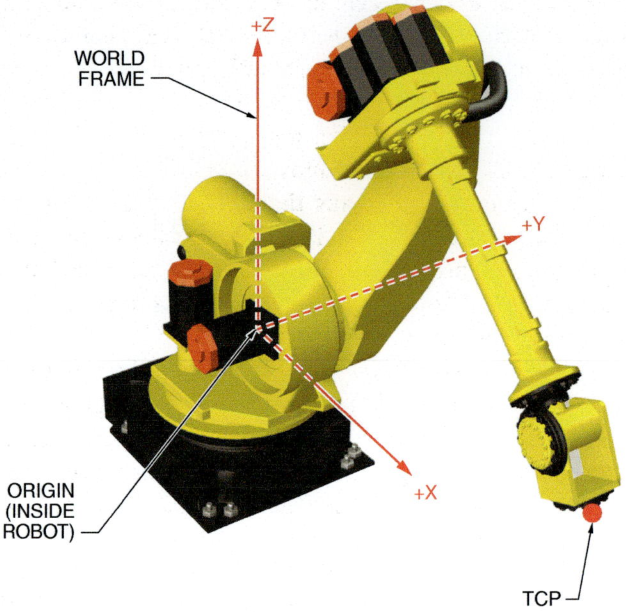

Figure 2-8. The predefined World frame is a Cartesian frame with its origin inside the robot and the axes in certain directions from the robot.

Right-Hand Rule

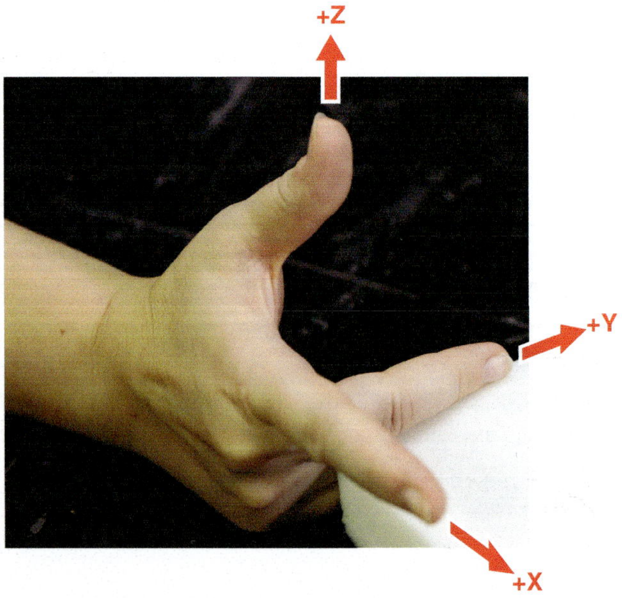

Figure 2-9. The right-hand rule is an easy way to remember the orientations of the X, Y, and Z axes in a Cartesian frame.

Singularity

There is a specific robot position that can be a problem when both jogging and running a program. If the robot Axes 4 and 5 are in a perfectly straight line, the software indicates a singularity fault. **See Figure 2-10.** *Singularity* is a specific robot position in which any subsequent movement can be accomplished in multiple ways and the robot controller cannot choose which one to execute.

Singularity only occurs during Cartesian jogging because this method relies on the controller to calculate the required movement of each joint based on the desired TCP movement. Singularity means there is more than one solution to move the robot from one point to another. Because the robot cannot make decisions on its own, the operator must control every aspect of the robot. To ensure that a singularity fault does not sound, Axes 4 and 5 must never be in alignment nor attempt to pass through alignment.

When a singularity fault sounds, it is not possible to jog the robot in Cartesian. The only way to escape this situation is to change to joint jogging and move the robot out of the position. The operator first resets the fault and changes the method to joint jogging. Then Axis 5 is jogged at least 10° in either direction, which is enough to disrupt the singularity condition. The robot can then be returned to a Cartesian jogging method to continue jogging.

FANUC has a software add-on available called Singularity Avoidance that prevents this situation. However, programmers have been working around this singularity issue for years. An experienced programmer is aware of this problem and can easily program around it, saving their company the added expense of purchasing this software.

Singularity

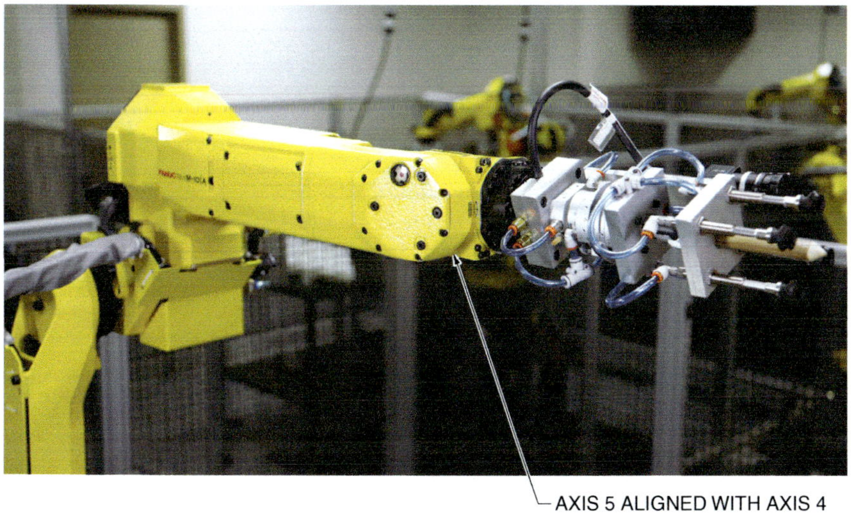

Figure 2-10. One specific robot position, with Axes 4 and 5 perfectly aligned, is known as singularity. If this position is entered while Cartesian jogging, the robot cannot move out of it without switching to joint jogging.

SECTION 2.3 JOGGING ROBOTS

Before a robot can be jogged, a number of conditions must be met. These allow the robot to be jogged while also ensuring the safety of the operator who may be within the robot cell. The necessary conditions are the following:

1. The key switch on the controller must be in the T1 position.
2. The teach pendant must be turned ON.
3. The deadman switch on the back of the teach pendant must be held in position 2, while the SHIFT button is pressed and held continuously. **See Figure 2-11.** The deadman switch has three positions. Either position 1 (not pressed) or position 3 (fully pressed) prevents the robot from moving. Holding the deadman switch in position 2 (halfway pressed) allows the robot to be jogged.
4. The robot must not be in a fault condition. Faults (also called alarms) are indicated in two places. On the controller, a red indicator is lit for a fault and unlit when ready to jog. On the teach pendant, a certain area of the display, labeled Fault, is red for a fault and green when ready to jog. **See Figure 2-12.** If the robot indicates a fault, read the fault message on the TP screen. Most likely, the fault message will show "Deadman switch released" because the deadman switch is not held in position 2. (It takes practice to get the feel of position 2 on the deadman switch and maintain that position while making other selections. It may help to support the weight of the teach pendant on a table while becoming familiar with the proper pressures required on the deadman switch and SHIFT button.) **See the *Resetting a Fault* Procedure.** If the fault message is something else, investigate other causes and procedures.

Deadman Switches

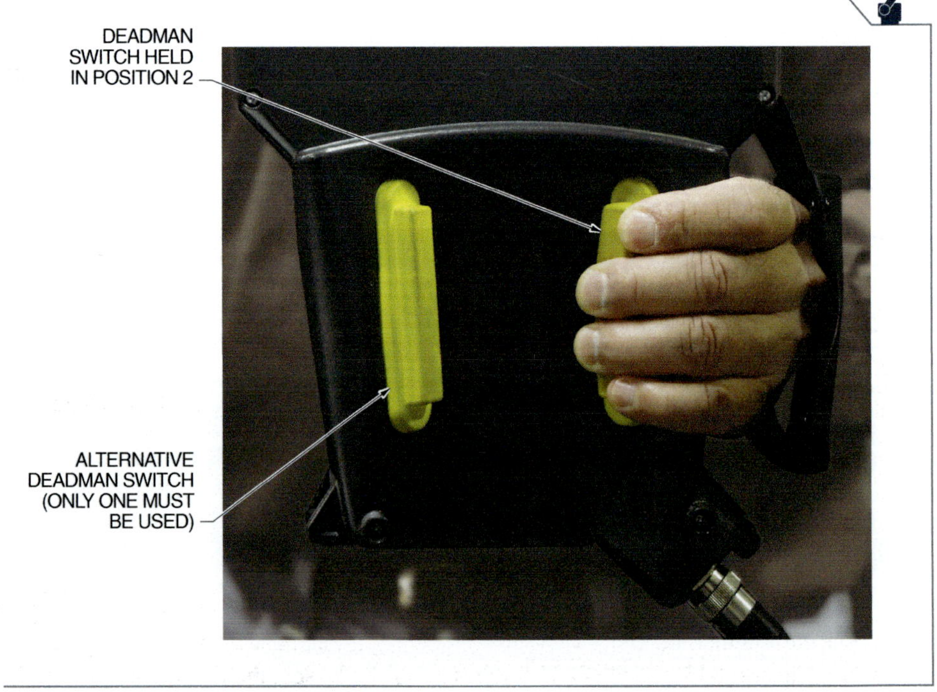

Figure 2-11. One of the deadman switches must be held in position 2 (halfway pressed) in order to jog the robot.

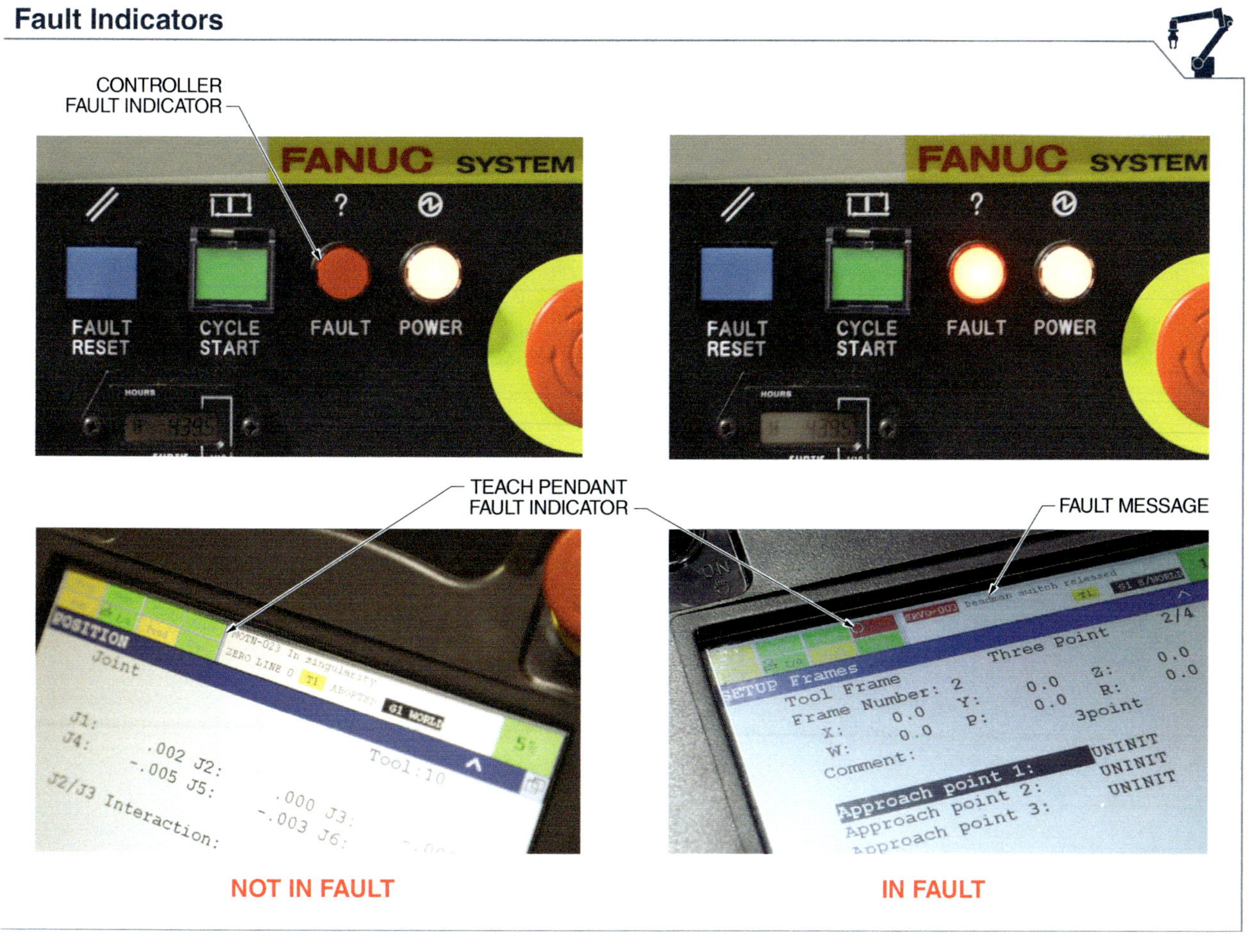

Figure 2-12. There is a fault light indicator on both the teach pendant display and the controller panel.

PROCEDURE

Resetting a Fault

1. Release all buttons.
2. Re-press and hold the deadman switch in position 2.
3. Press SHIFT and then press and release RESET.
4. If the fault continues, the condition causing the fault is not resolved. Investigate further actions.

Selecting the Jogging Method

When selecting the method of jogging a robot, the operator presses the COORD button. The current method is shown near the upper right-hand corner of the teach pendant screen. **See Figure 2-13.** As COORD is pressed and released, the five jogging methods (joint, JGFRM, World, tool, and user) will cycle through. The operator continues pressing COORD until the desired method of jogging appears.

There is a slight delay when the button is pressed and released. If COORD is pressed too rapidly, the desired jog method can easily be passed. If this happens, the operator can simply continue pressing COORD until the desired jogging method reappears.

Jog Settings

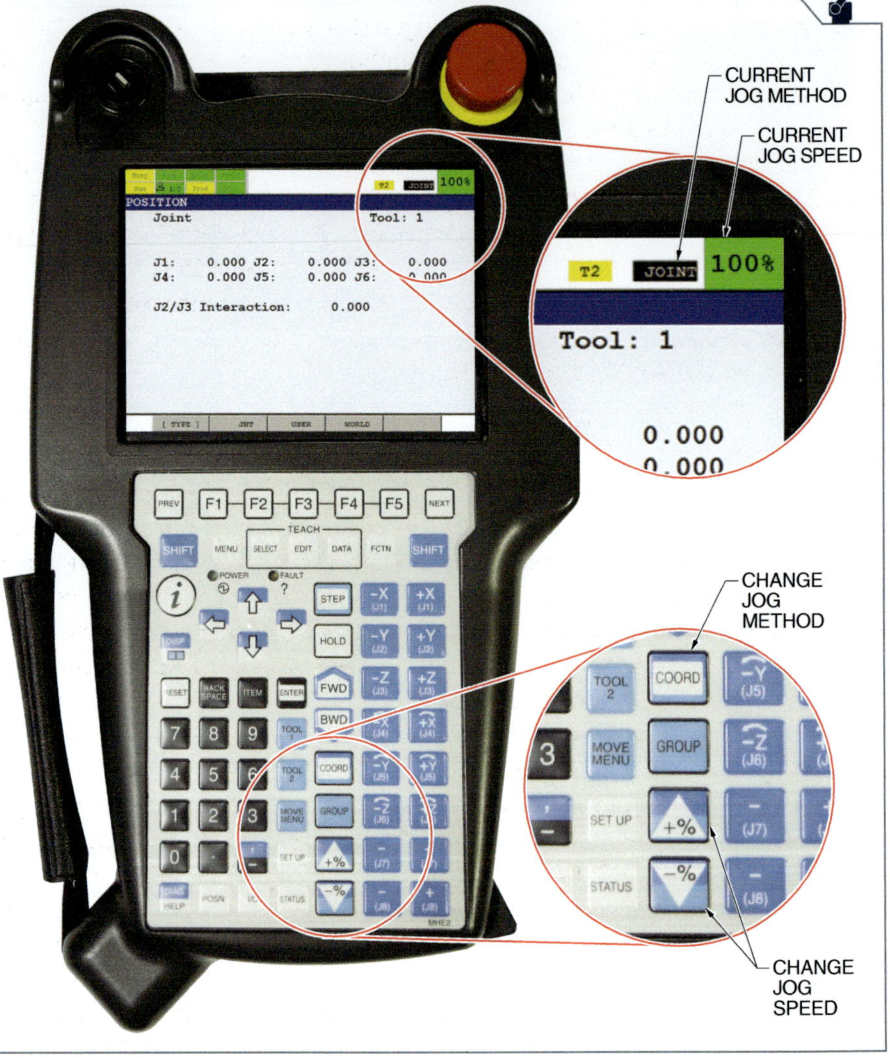

Figure 2-13. Pressing the COORD button cycles through the five jogging methods. A set of up- and down-arrow buttons are used to change the jogging speed. The current settings are shown on the display.

Adjusting Jogging Speed

When jogging, there is always a potential to crash the robot, which could damage the end effector, robot, and/or surrounding equipment. It is important to slow the robot down considerably when near equipment and know the direction the robot will move when the jog buttons are pressed.

A pair of speed buttons adjusts the robot's speed. Pressing the △ button increases the robot's speed, and the ▽ button reduces its speed.

The currently selected speed is shown in the upper right-hand corner of the teach pendant screen. The arrow buttons are half blue and half white. This indicates that the operation of these buttons is different when SHIFT is pressed or not pressed. If SHIFT is pressed and held, each press and release of the up-arrow button results in increments from very fine, fine, 5%, 50%, and 100%. If SHIFT is not held, the speed changes in smaller increments from very fine, fine, 1%, 2%, 3%, 4%, 5%, 10%, 15%, 20%, and so on to 100%.

When the speed very fine or fine is selected, the robot moves in tiny increments. A speed of very fine moves an axis 0.0001° (in joint) or the TCP 0.002 mm (in Cartesian). A speed of fine moves an axis 0.001° (in joint) or the TCP 0.023 mm (in Cartesian). This means that each time a jog button is pressed, the robot only moves that amount. To move the robot further, a jog button must be pressed again.

The percentage jog speed settings are referenced to the maximum jog speed, which is 10% of the robot's full speed. This means that if the robot's full speed is 2500 mm/sec, the maximum jog speed is 250 mm/sec. When the TP displays 100%, the TCP's actual speed is 250 mm/sec. If the TP is set at 10%, the TCP speed is 25 mm/sec.

Also, the robot's system variables can be altered to prevent the robot from setting jog speeds above 50%. The robot's operator needs to be aware of any altered variables when operating a robot they are unfamiliar with.

Axis Limits

All robots have axis limits to prevent over-rotation. An *axis limit* is a rotational position beyond which a joint cannot rotate, either by programmed or physical limits. Rotation is limited by the electrical connections that must pass through each joint to the next joints and the end effector. If the robot fault message shows an Axis Limit or Position Not Reachable fault, one or more of the robot's axis limits have been reached. It may not be obvious which axis caused the fault, so operators must first determine this by either checking the Position screen or reading the fault code.

On the teach pendant, operators must press POSN and then press F2 `JNT` to show the Position screen. **See Figure 2-14.** This screen shows the position of each axis in degrees. The axis readings are compared with the System Axis Limits screen. The System Axis Limits screen is accessed by pressing MENU, then NEXT, and then SYSTEM. The operator can then select `Axis Limits` and press ENTER. The axis causing the fault is the one with a position (in degrees) matching or very close to the axis limit. It may be helpful to use the double-screen feature of the display when comparing.

Alternatively, the Axis Limit fault provides a code that indicates the axis or axes involved. If an operator learns to read the code, it is a shortcut to resolving the alarm that avoids having to check the Position screen. An example code is G1A4LA3N. The "G1" means Group 1, which is the robot. Axes are identified with an "A" followed by the axis number and either an "N" (for near limit) or an "L" (for at limit). Therefore, in this example, Axis 4 is at limit, and Axis 3 is near limit.

To stop the fault, the applicable axis is jogged away from the axis limit. To avoid the fault when running a program, one or more positions must be reprogrammed to prevent the robot from trying to reach a position beyond the axis limits.

When learning about jogging and axis limits, it is helpful to practice putting the robot into an all-zero position. The operator moves each axis until it reaches 0. To do this, the operator selects the joint jog method by pressing and releasing COORD. Then, the operator starts with Axis 1 and jogs the robot until it reaches 0. The remaining axes are jogged in sequential order until all axes are at 0. When completed, Axis 5 is jogged in the negative direction at least 10°. This ensures the robot does not sound a singularity fault when jogging in the Cartesian method.

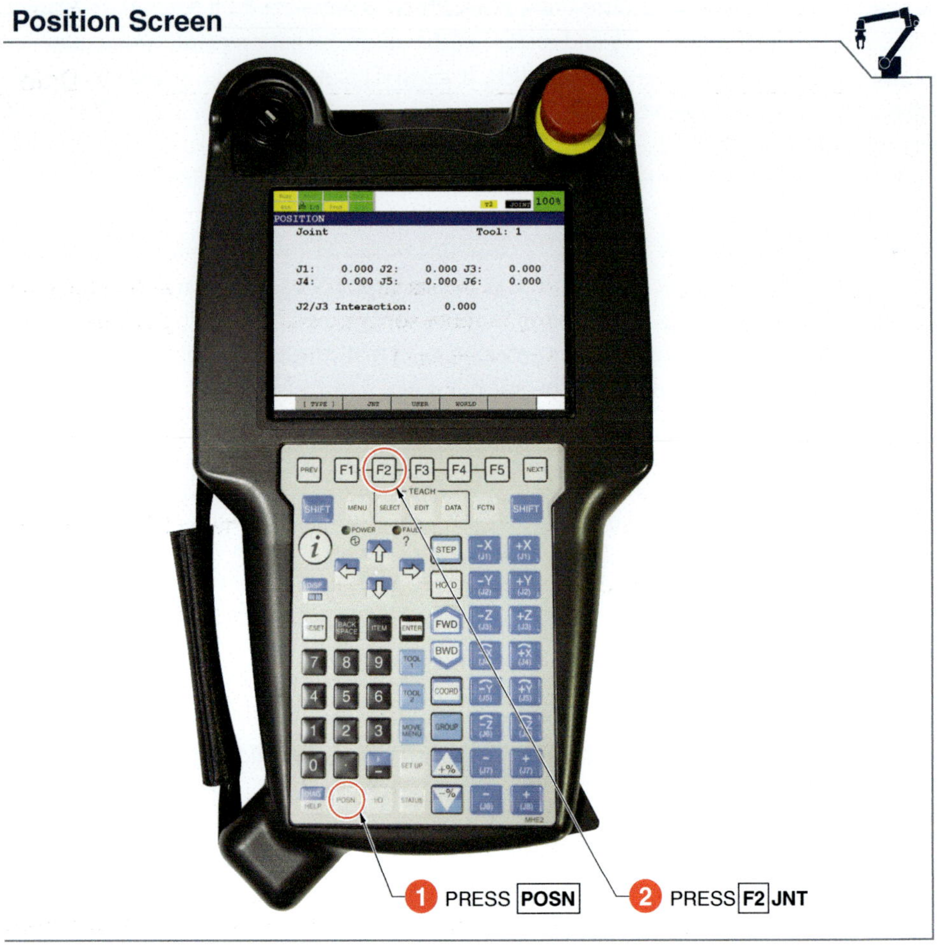

Figure 2-14. The Position screen shows the current position of the robot and each axis in degrees.

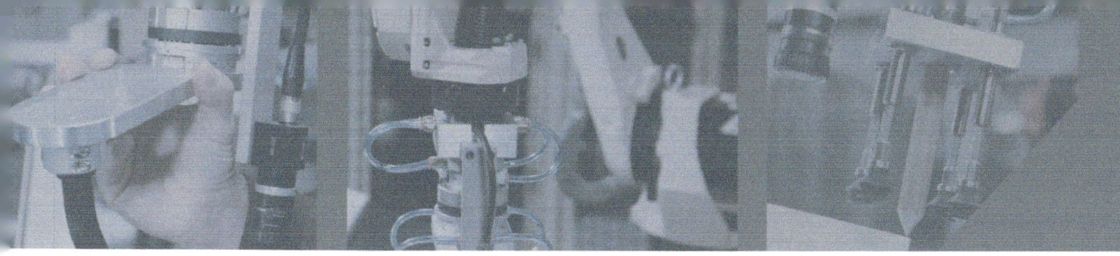

Review 2

Name ______________________ Date ____________

SECTION 2.1 POWERING UP ROBOTS

_______ **1.** ___ start is the startup procedure for powering up a robot in preparation for adding or modifying robot software and overriding errors.

A. Coordinated
B. Controlled
C. Hot
D. Cold

_______ **2.** Jogging requires that the teach pendant ON/OFF switch be in the ___ position.

3. Describe the standard method for starting up a robot.

SECTION 2.2 COMPARING JOGGING METHODS

_______ **4.** ___ is the manual adjustment of a robot's position and orientation via the teach pendant.

_______ **5.** A(n) ___ axis is one of the first three axes (1, 2, or 3).

A. primary
B. main
C. major
D. initial

_______ **6.** Joint jogging is generally used for ___ movements.

_______ **7.** ___ jogging is a jogging method in which all axes operate simultaneously (as needed) in order to move the tool center point (TCP) in a straight line or to move the robot about the TCP.

A. Cartesian
B. Joint
C. Frame
D. Singularity

_______ **8.** Jogging the robot in reference to a(n) ___ is known as Cartesian jogging.

____________________ **9.** The four types of Cartesian frames to choose from when jogging a robot are World, tool, user, and ___.

A. joint
B. program
C. cell
D. jog

____________________ **10.** The only frame preset by the manufacturer is ___.

____________________ **11.** Singularity only occurs during ___ jogging.

12. Which buttons are used to rotate individual axes when joint jogging?

13. Since Axis 2 is interactive with Axis 3, what happens when Axis 2 is jogged in joint?

14. What is the purpose of the J7 and J8 buttons?

15. When Cartesian jogging, how does the TCP move when a major axis button is pressed?

16. When Cartesian jogging, how does the TCP move when a minor axis button is pressed?

17. What is the right-hand rule?

18. What is singularity?

19. How is singularity escaped?

SECTION 2.3 JOGGING ROBOTS

________________ **20.** Before jogging, the deadman switch must ___.

A. not be pressed
B. be fully pressed
C. be pressed halfway
D. be disabled

________________ **21.** Before jogging, the key switch must be in the ___ position.

________________ **22.** When selecting the method of jogging a robot, the operator presses the ___ button.

________________ **23.** The currently selected jog speed is shown in the ___ of the teach pendant screen.

A. upper right-hand corner
B. lower left-hand corner
C. function button area
D. speed menu area

24. What does the SHIFT button do while selecting jog speed with the arrow buttons?

25. What is an axis limit?

2 Lab Exercise

Lab Objective

Practice approaching and establishing positions with a robot at certain locations on separate equipment without actual physical contact.

Equipment Required

- Robot with a gripper-style end effector
- Pointing tool, such as a pen, marker, or pointed dowel
- Square object, placed in the robot cell

Lab Procedure

1. Place the pen or marker in the robot end effector. The pen or marker may need to be taped in the end effector to hold it steady.
2. Place the small square object in the robot cell at an angle to the robot.
3. While in joint, jog the robot to each corner of the square object, lightly touching the pen or marker to each corner. Use a piece of paper as a feeler gauge by sliding the paper between the pen or marker and the object. Use the speed buttons and reduce the robot speed to ensure the robot does not put pressure between the pen or marker and the object. The marker should be vertical and at a right angle to the square.
4. Change the jog method to World. Again, jog the robot to each corner of the square object using paper as a feeler gauge to ensure pressure is not applied to the marker or pen. Use the robot speed buttons appropriately, speeding up slightly when the robot is not near an object and slowing down when a robot approaches an object. Use fine and very fine when very close to the paper. Remember these two speeds are incremental.

3 Robot Frames

FANUC robots use four types of robot frames. The manufacturer creates one frame, while the other three must be created by the robot programmer as needed. Each frame is used in different situations to assist the programmer when creating robot programs and when recovering from faults that may arise during the robot's operation. Properly using these frames makes a precise and robust program.

Objectives

SECTION 3.1 REFERENCING FRAMES

- Compare the two purposes and procedures for referencing frames.
- Identify the location of the origin for the World frame.

SECTION 3.2 CREATING TOOL FRAMES

- Describe the purpose for creating a tool frame TCP.
- Demonstrate the direct-entry method of creating a tool frame TCP.
- Demonstrate the three-point method of creating a tool frame TCP.
- Demonstrate the six-point (XZ) method of creating a tool frame TCP.
- Explain how to test a tool frame.

SECTION 3.3 CREATING USER FRAMES

- Describe the purpose for creating a user frame.
- Demonstrate the three-point method of creating a user frame.
- Explain how the four-point method differs from the three-point method.
- Demonstrate the direct-entry method of creating a user frame.
- Explain how to test a user frame.

SECTION 3.4 CREATING JOG FRAMES

- Describe the purpose for creating a jog frame.
- List the two methods of creating jog frames.

Key Terms

- tool frame
- user frame
- jog frame

Learner Resources
atplearningresources.com
Access Code: **762316**

SECTION 3.1 REFERENCING FRAMES

A robot can use different types of frames to define points in space and its interaction with them. There are two ways in which a frame is referenced when operating or programming a robot. The distinction can be confusing at first, but ease will come with practice.

First, a frame is chosen for the robot jogging method by pressing the COORD button. Pressing it repeatedly cycles through the list of choices: Joint, JGFRM (jog frame), World, Tool, and User. The currently selected jogging method appears near the top-right corner of the teach pendant screen. Selecting a jogging method/frame is only for moving the robot into desired positions and does not affect the definition of points in a program.

Alternatively, frames are chosen for programming in the Jog menu, which is accessed by pressing and holding SHIFT and then pressing COORD. A small menu appears with Tool, Jog, and User frame settings, each with a number for the currently selected frame. **See Figure 3-1.** The programmer selects the desired frame type and changes the frame number as needed. These frame settings are used when adding position points to a program or when setting up certain other frames. For example, a point in space can be referenced to Tool Frame 1 and User Frame 1. Another point may be referenced to Tool Frame 1 and User Frame 2. Also, when creating a user frame, the correct tool frame number must be selected in the Jog menu.

Jog Menu

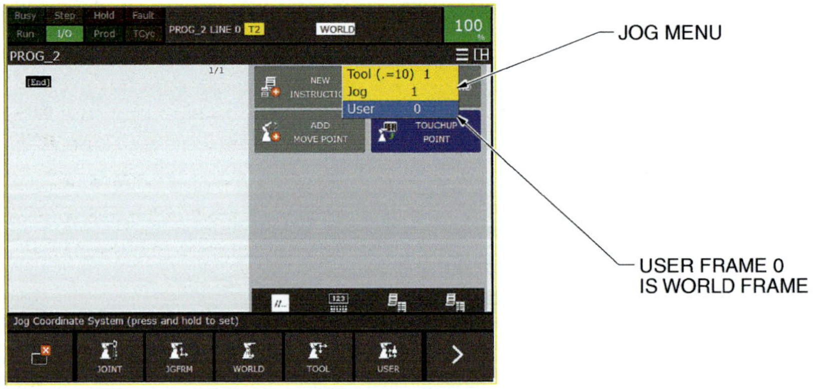

Figure 3-1. Tool, jog, and user frames are chosen in the Jog menu.

The World frame is the only robot frame created by the manufacturer and immediately ready to use. For the World frame, the origin is centered at the base of Axis 1. More precisely, if an imaginary vertical line were drawn through the center of the Axis 1 gear shaft and an imaginary horizontal line were drawn through the Axis 2 motor shaft, the origin of the World frame would be where the two lines intersect. **See Figure 3-2.** When the operator programs in the World frame, each of the programmed points references the World frame and its origin. For example, a point programmed with the coordinates of 5, 4, 2 is 5 mm from the origin in the +X direction, 4 mm from the origin in the +Y direction, and 2 mm from the origin in the +Z direction. In the Jog menu, User Frame 0 represents the World frame.

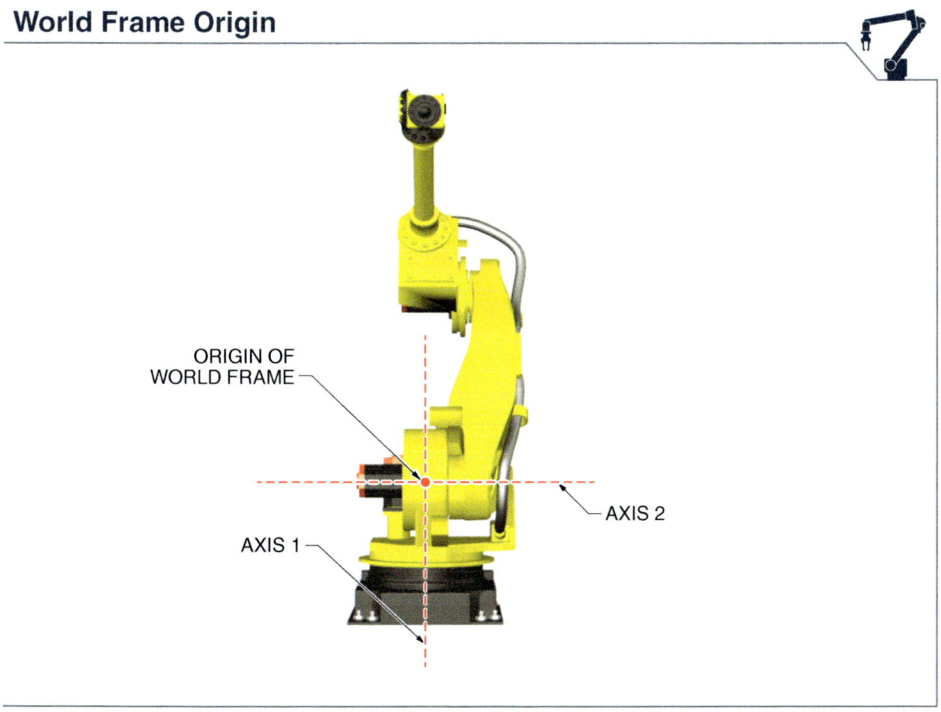

Figure 3-2. The origin of the World frame is at the intersection of Axis 1 and Axis 2.

SECTION 3.2 CREATING TOOL FRAMES

A *tool frame* is a robot frame with an origin at the tool center point (TCP) and axes that extend from that point in three directions (X, Y, and Z). **See Figure 3-3.** The TCP is a point centered on the end effector and where the work is done. Tool frames are needed because there are many different types of end effectors, and each has a specific size and shape as well as specific points at which it interacts with a product. The robot must be taught each end effector's point of reference, which is its TCP.

It is common for one robot to require different end effectors to complete a variety of tasks. A robot may have the capability to release an end effector and pick up a different one. Because of this capability, a robot can have up to 10 different tool frames. Each tool frame is then used for a different end effector. **See Figure 3-4.** During a program, end effectors can be automatically changed, or one end effector can have multiple uses, and each would require a unique tool frame.

Once an end effector is attached to a robot, the factory-established TCP at the end of Axis 6 is of little use to the programmer. The programmer must re-establish the TCP to conform to the end-effector location and orientation. If this is not done, the robot could crash into another piece of equipment in the work cell and damage the end effector. Also, repaired end effectors are essentially impossible to realign to their exact previous position, so the tool frame must be re-established.

Establishing a tool frame is essentially just teaching the robot where the TCP of the end effector is in space. There are five methods to choose from when establishing a tool frame. The three common methods are the direct-entry, three-point, and six-point (XZ) methods. The other two methods are the six-point XY and two-point +Z methods. These last two are uncommon and generally used for very specific applications.

Tool Frames

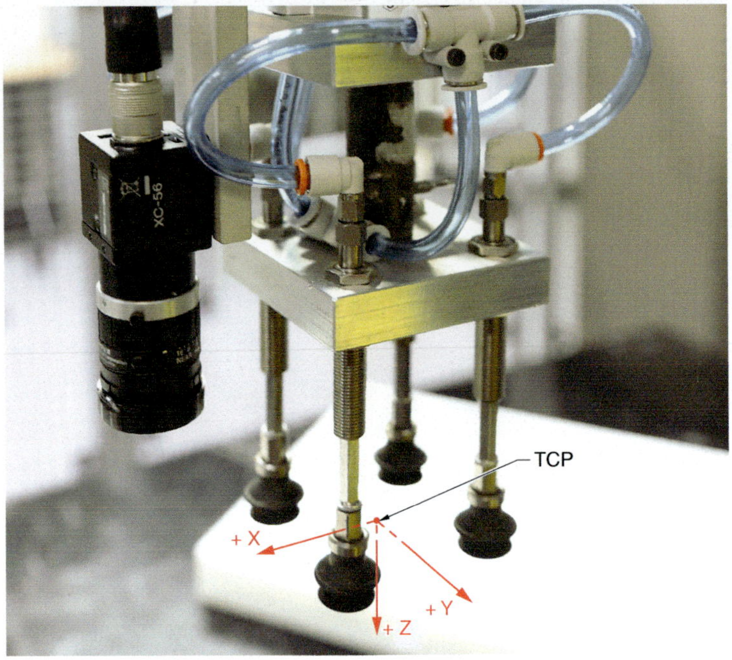

Figure 3-3. A tool frame specifies the TCP and axis directions for a particular end effector.

End Effector TCPs

Figure 3-4. Each end effector must have its own tool frame.

Using the Direct-Entry Method to Establish Tool Frames

To establish a tool frame using the direct-entry method, the programmer must have a schematic of the end effector with all dimensions. **See Figure 3-5.** The end effector must also be new—never before used—so that it is certain that it has never been damaged. A damaged end effector, even if repaired, may not match the original dimensions on the schematic.

End-Effector Schematics

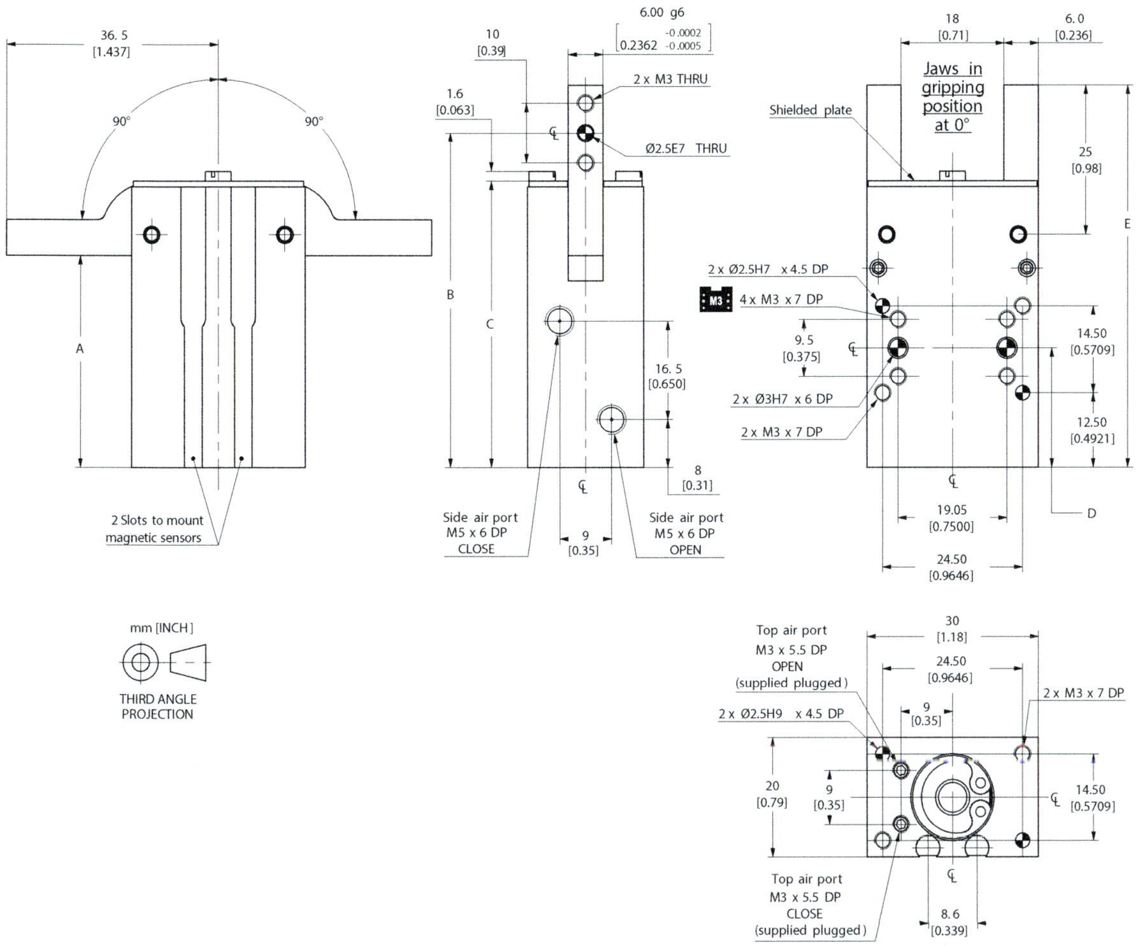

Figure 3-5. The direct-entry method for creating a tool frame can only be used if there is a schematic of the end effector.

An end-effector schematic shows the mounting of the end effector and the location and orientation of its TCP with respect to the center of Axis 6 on the robot. The location information includes the distances in the X, Y, and Z directions, and the orientation information includes the angles in the W (yaw, or around Z), P (pitch, or around Y), and R (roll, or around X) directions. The programmer enters these dimensions into the robot controller. An offset in the Z direction means that the end effector is centered in the X and Y directions but is extended away from the robot arm in a straight line. **See the *Establishing a Tool Frame Using the Direct-Entry Method* Procedure.**

PROCEDURE

Establishing a Tool Frame Using the Direct-Entry Method

Choose the Tool Frame to Set Up

1. On the teach pendant, press the MENU button.
2. Use the arrow buttons to select `SETUP`.
3. Use the arrow buttons to move to the right, select `Frames`, and press ENTER.

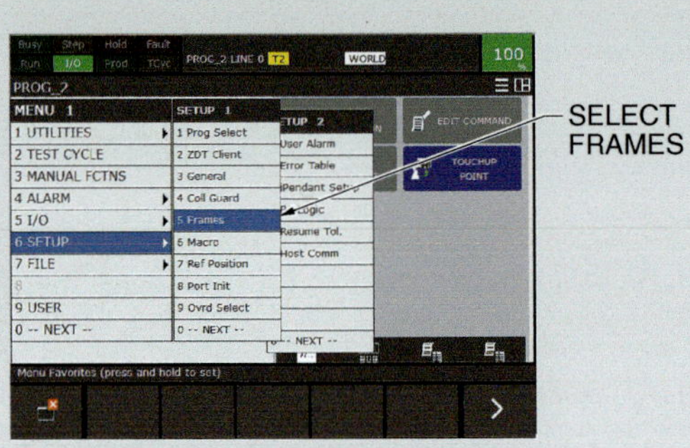

4. Verify that the top-left of the display shows `Tool Frame`. If it does not, then change the frame type:
   a. Press F3 `[OTHER]` to show the `OTHER` menu.
   b. Use the arrow buttons to select `Tool Frame` and press ENTER.

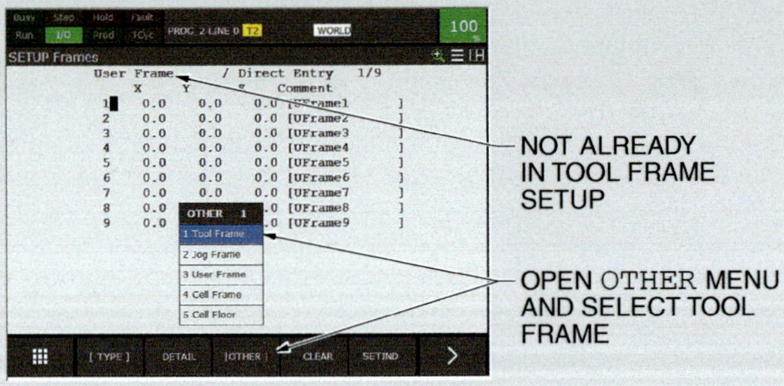

5. Use the arrow buttons to select the desired tool frame to establish.

Choose Method

6. Press F2 `DETAIL` to enter the `SETUP Frames` screen.
7. Press F2 `[METHOD]` to show the `METHOD` menu.

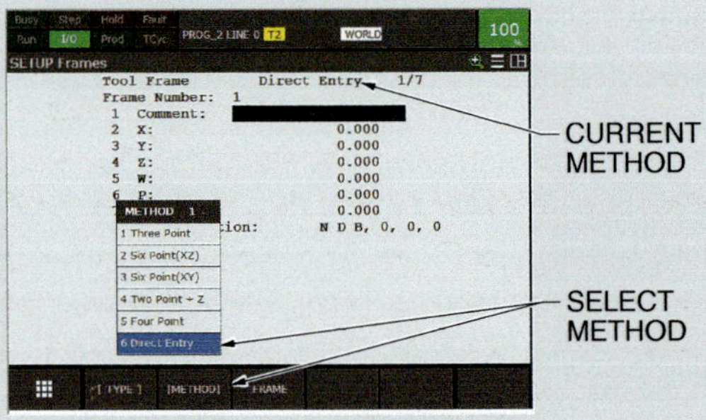

8. Use the arrow buttons to select `Direct Entry` and press ENTER.

Name the Tool Frame

⑨ Use the arrow buttons to select the `Comment` field and press ENTER.

⑩ Use the arrow buttons to select `Options/Keybd`.

⑪ Press F5 `KEYBOARD`.

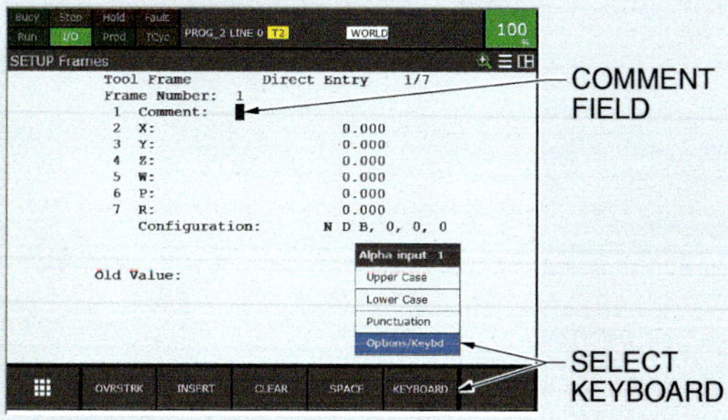

⑫ Using the teach pendant keyboard, enter a comment or name for the tool frame and press EXIT. (Remember that no spaces or unique characters can be used.)

Enter Dimensions

⑬ Use the arrow buttons to select each direction or orientation offset line (as needed) and press ENTER.

⑭ Enter the associated dimension (in millimeters) from the schematic and press ENTER.

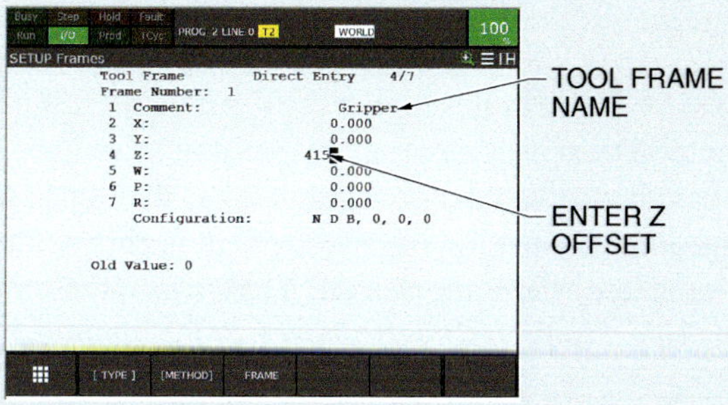

⑮ When completed, press PREV to return to the `SETUP Frames` screen.

Using the Three-Point Method to Establish Tool Frames

The three-point method can be used with most end effectors that are in a straight line with the robot's Z direction (straight out from the end of the robot). Creating a three-point tool frame requires two pointers, one on the end effector and one fixed to a work area, to be lined up three times. **See Figure 3-6.** The robot pointer must be centered in the X and Y directions of the end effector and slightly past the end of the end effector. The fixed pointer can be located anywhere within the robot cell, so long as the robot can approach the fixed pointer from all directions without putting the robot in singularity.

In industry, both pointers are made of metal and sharpened to fine points for highly accurate alignment. The pointers may be magnetically adhered or threaded to the robot and work surface. In an educational setting, using a wooden dowel as a robot pointer works well. The wooden pointer does not cause serious damage to the robot or end effector if a light crash occurs. The dowel can then be sharpened as needed if it is accidentally pressed into an object and dulled. The fixed pointer can also be made of a material that can deform or break before causing damage to the robot.

Three-Point Method for Tool Frames

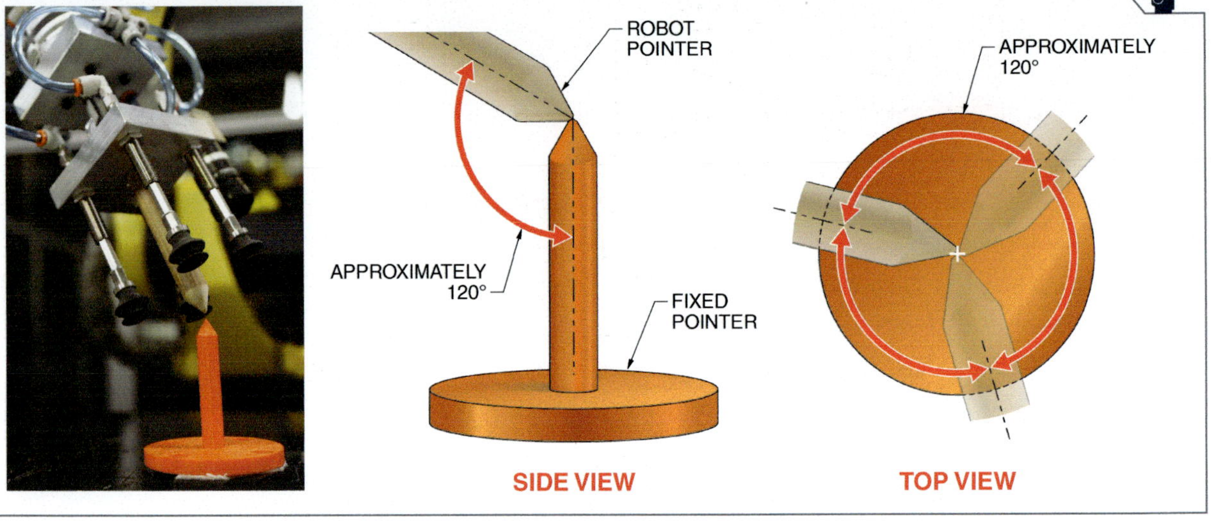

Figure 3-6. In the three-point method, a robot pointer is aligned with a fixed point three times, each from a different direction.

Navigating to the SETUP Frames screen is the same as for the direct-entry method. The three-point method then uses an arbitrary point in space (represented by the fixed pointer) and has the robot record the position of its TCP at this point three times from three different directions. By having the robot orient to this point from three different directions, it can calculate where the desired TCP is in three-dimensional space. **See the *Establishing a Tool Frame Using the Three-Point Method* Procedure.**

PROCEDURE

Establishing a Tool Frame Using the Three-Point Method

Choose the Tool Frame to Set Up

1. On the teach pendant, press the [MENU] button.
2. Use the arrow buttons to select `SETUP`.
3. Use the arrow buttons to move to the right, select `Frames`, and press [ENTER].
4. Verify that the top-left of the display shows `Tool Frame`. If it does not, then change the frame type:
   a. Press [F3] `[OTHER]` to show the `OTHER` menu.
   b. Use the arrow buttons to select `Tool Frame` and press [ENTER].
5. Use the arrow buttons to select the desired tool frame to establish.

Choose Method

6. Press [F2] `DETAIL` to enter the `SETUP Frames` screen.

⑦ Press F2 `[METHOD]` to show the `METHOD` menu.

⑧ Use the arrow buttons to select `Three Point` and press ENTER.

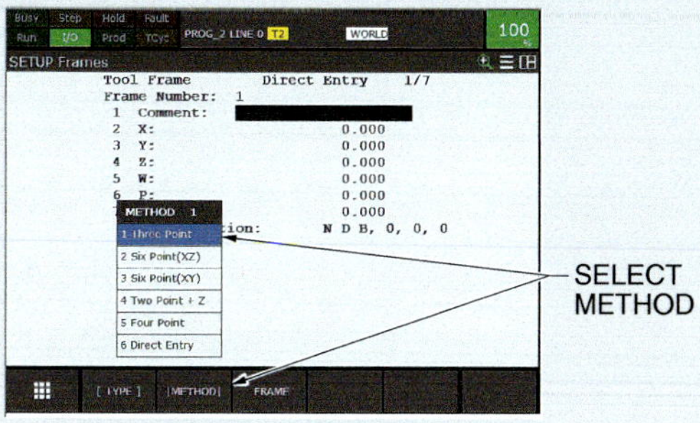

Name the Tool Frame

⑨ Use the arrow buttons to select the `Comment` field and press ENTER.

⑩ Use the arrow buttons to select `Options/Keybd`.

⑪ Press F5 `KEYBOARD`.

⑫ Using the teach pendant keyboard, enter a comment or name for the tool frame and press EXIT. (Remember that no spaces or unique characters can be used.)

Record the TCP Position

⑬ Use the arrow buttons to select `Approach point 1`.

⑭ Jog the robot in World to line up the end-effector pointer with the fixed pointer.

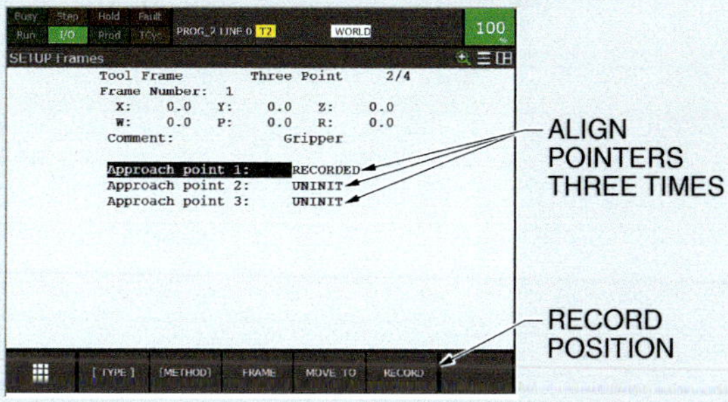

The pointer of the robot should be at an approximate 120° angle to the fixed pointer and just touching.

⑮ Press and hold SHIFT and then F5 `RECORD`.

⑯ Use the arrow buttons to select `Approach point 2`.

⑰ Jog the robot to line up the end-effector pointer with the fixed pointer, again with approximately 120° between the pointers. However, this time, also align the end-effector pointer at approximately 120° horizontally from the first orientation.

⑱ Press and hold SHIFT and then F5 `RECORD`.

⑲ Use the arrow buttons to select `Approach point 3`.

⑳ Jog the robot to line up the end-effector pointer with the fixed pointer. This point is located at a 120° angle from the fixed pointer and 120° from both the first and second approach points.

㉑ Press and hold SHIFT and then F5 `RECORD`.

㉒ Press PREV to return to the `SETUP Frames` screen.

Using the Six-Point (XZ) Method to Establish Tool Frames

The six-point (XZ) tool frame must be used when the end effector's TCP is not close to zero in the X direction as well as the Z direction. A welding torch is a perfect example of this type of end effector, which is angled in a way to reach tight corners. **See Figure 3-7.**

Offset End Effector

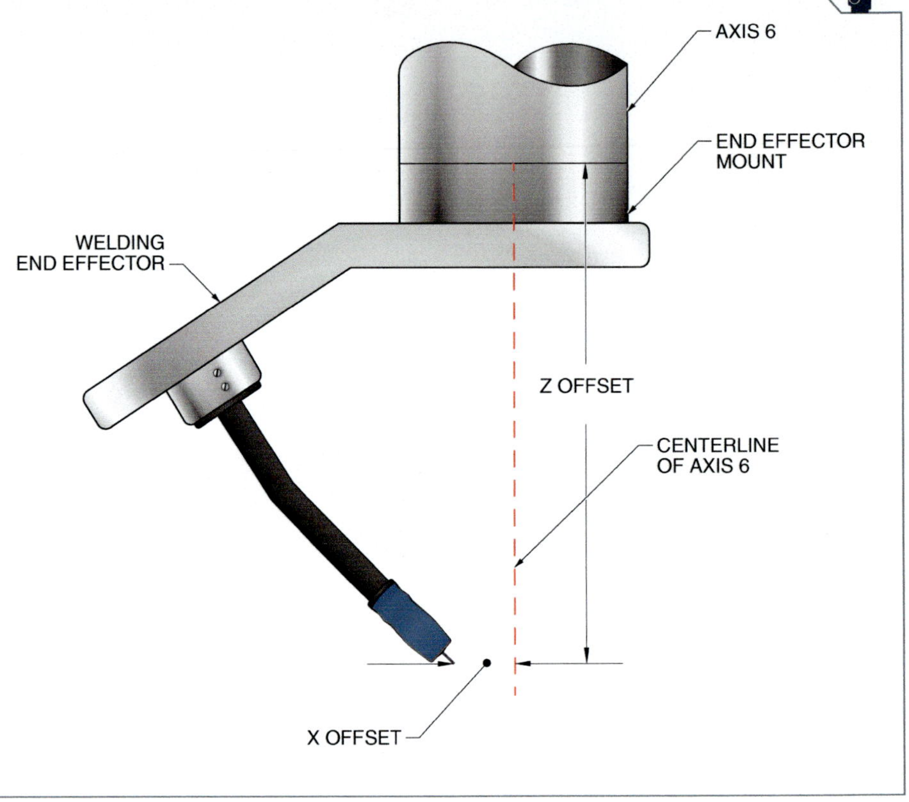

Figure 3-7. The six-point (XZ) method is commonly used for end effectors that are offset in both the X and Z directions.

For educational purposes, a 3D-printed simulation weld torch may be used with the same fixed pointer used in the three-point method. This is for the protection of the robot if a student were to crash it. The fixed pointer can be located anywhere in the robot cell as long as it can be reached on all sides by the robot without putting it in singularity.

The six-point (XZ) method is much the same as the three-point method. However, after teaching the robot the location of the TCP with the approach points, this method adds extra steps to teach the robot which directions will be considered X and Z. **See Figure 3-8.** This is important for the robot to be able to orient correctly for an offset end effector. **See the *Establishing a Tool Frame Using the Six-Point (XZ) Method* Procedure.**

Six-Point (XZ) Method

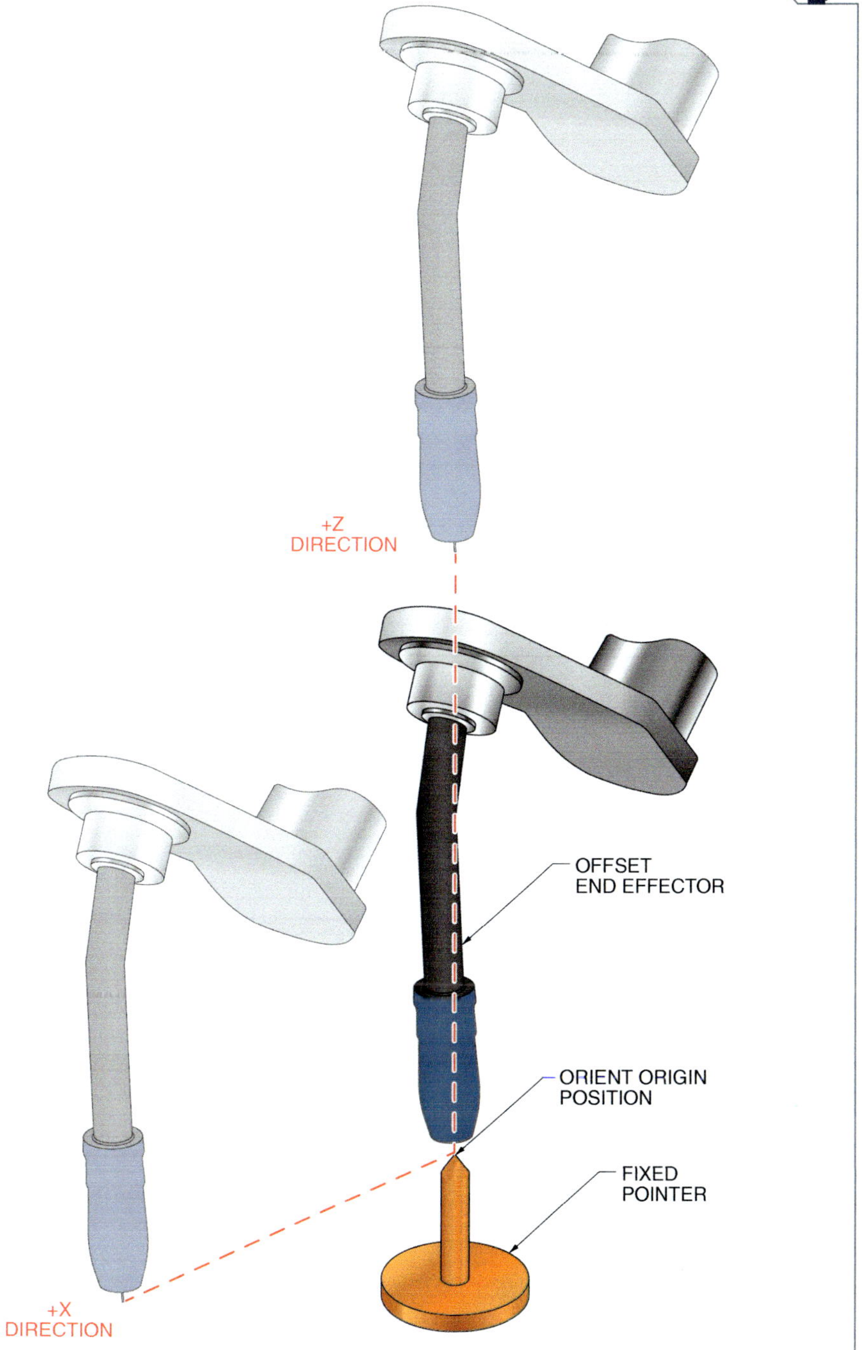

Figure 3-8. The six-point (XZ) method begins the same way as the three-point method, but it adds recorded points in the X and Z directions in order to define the end-effector orientation.

PROCEDURE

Establishing a Tool Frame Using the Six-Point (XZ) Method

Choose the Tool Frame to Set Up

1. On the teach pendant, press the MENU button.
2. Use the arrow buttons to select SETUP.
3. Use the arrow buttons to move to the right, select Frames, and press ENTER.
4. Verify that the top-left of the display shows Tool Frame. If it does not, then change the frame type:
   a. Press F3 [OTHER] to show the OTHER menu.
   b. Use the arrow buttons to select Tool Frame and press ENTER.
5. Use the arrow buttons to select the desired tool frame to establish.

Choose Method

6. Press F2 DETAIL to enter the SETUP Frames screen.
7. Press F2 [METHOD] to show the METHOD menu.
8. Use the arrow buttons to select Six Point (XZ) and press ENTER.

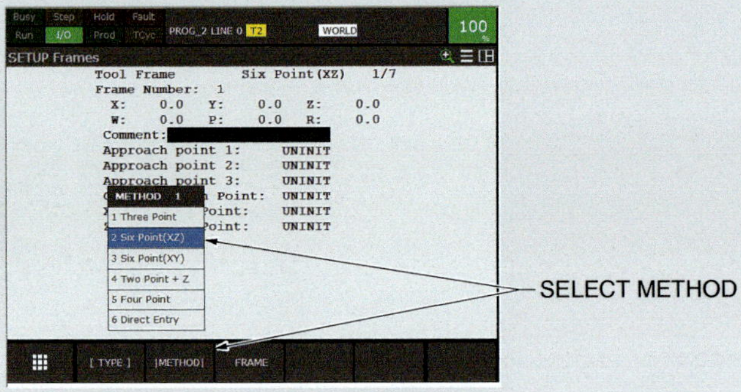

Name the Tool Frame

9. Use the arrow buttons to select the Comment field and press ENTER.
10. Use the arrow buttons to select Options/Keybd.
11. Press F5 KEYBOARD.
12. Using the teach pendant keyboard, enter a comment or name for the tool frame and press EXIT. (Remember that no spaces or unique characters can be used.)

Record the TCP Position

13. Use the arrow buttons to select Approach point 1.
14. Jog the robot in World to line up the end-effector pointer with the fixed pointer. The pointer of the robot should be at an approximate 120° angle to the fixed pointer and just touching.
15. Press and hold SHIFT and then F5 RECORD.
16. Use the arrow buttons to select Approach point 2.

17. Jog the robot to line up the end-effector pointer with the fixed pointer, again with approximately 120° between the pointers. However, this time, also align the end-effector pointer at approximately 120° horizontally from the first orientation.
18. Press and hold SHIFT and then F5 `RECORD`.
19. Use the arrow buttons to select `Approach point 3`.
20. Jog the robot to line up the end-effector pointer with the fixed pointer. This point is located at a 120° angle from the fixed pointer and 120° from both the first and second approach points.
21. Press and hold SHIFT and then F5 `RECORD`.

Record the TCP Orientation

22. Use the arrow buttons to select `Orient Origin Point`.

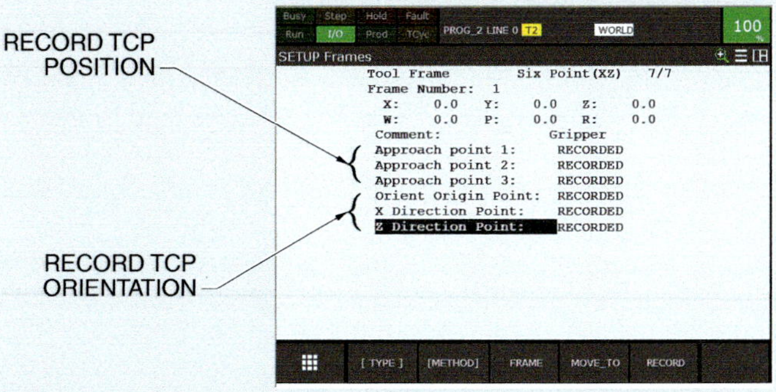

23. Jog the robot to line up the tip of the simulation weld torch at a 90° angle to the fixed pointer.
24. Press and hold SHIFT and then F5 `RECORD`.
25. Use the arrow buttons to select `X Direction Point`.
26. In World, jog the robot in the +X direction approximately 300 mm (12″).
27. Press and hold SHIFT and then F5 `RECORD`.
28. Use the arrow buttons to select `Orient Origin Point`.
29. Press F4 `MOVE_TO`. The robot moves in the X direction back to the origin point.
30. Use the arrow buttons to select `Z Direction Point`.
31. In World, jog the robot in the +Z direction approximately 300 mm (12″).
32. Press and hold SHIFT and then F5 `RECORD`.
33. Press PREV to return to the `SETUP Frames` screen.

Testing Tool Frames

A tool frame should be tested to ensure that it has been defined properly and that the end effector reacts as the programmer desires. A tool frame is tested by closely aligning the robot pointer and the fixed point, then jogging Axis 4 or 5, and then observing if the pointers remain aligned. **See the *Testing a Tool Frame* Procedure.**

PROCEDURE

Testing a Tool Frame

1. Jog the robot so that the end effector's TCP is aligned with, but approximately 0.25″ above, the fixed pointer.
2. On the teach pendant, press and hold the SHIFT button and then press COORD to show the Jog menu.
3. Use the arrow buttons to select `Tool`.
4. Press the number of the tool frame to be tested. The Jog menu then automatically disappears.
5. Select World jogging and set the speed to 10% or less.
6. Press the Axis 4 or Axis 5 jogging buttons in either direction.

If the tool frame has been created properly, the TCP of the end effector will remain in line with the fixed pointer as the robot arm moves around it. **See Figure 3-9.** If the pointer moves off the fixed pointer, the tool frame is not correct and must be re-created. It may take many attempts to produce a quality tool frame. Speed and accuracy come with time and experience.

Testing Tool Frames

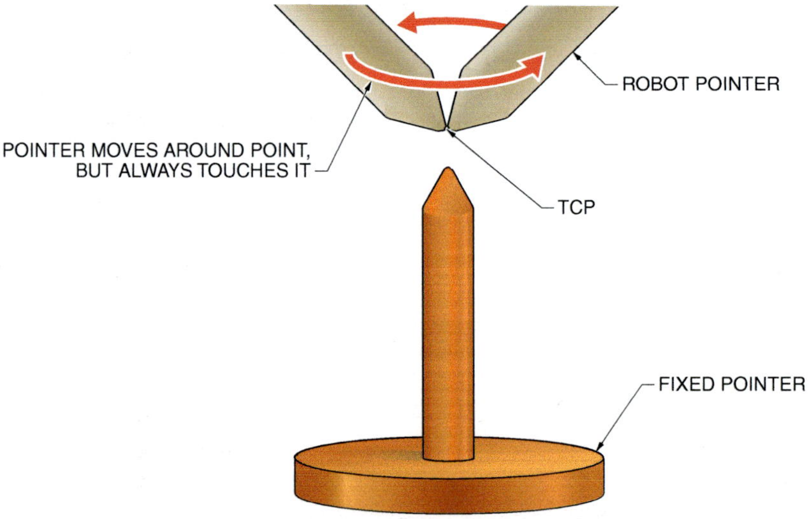

Figure 3-9. To test a tool frame, the robot pointer and fixed pointer are closely aligned and the robot's Axis 4 or Axis 5 is jogged around the point.

SECTION 3.3 CREATING USER FRAMES

A *user frame* is a robot frame with a user-defined origin and X, Y, and Z-axis directions, which is used to orient a robot to a particular object within a robot cell, such as a fixture. **See Figure 3-10.** User frames are created on other items located within a robot cell, such as fixtures, molds, and conveyors, with which the robot must interact. The origin may be defined at any convenient part of the object but is usually a corner. After a user frame is created, the programmer can initiate the user frame in the program. Then all points referenced to that user frame are based off that user-defined origin.

User Frames

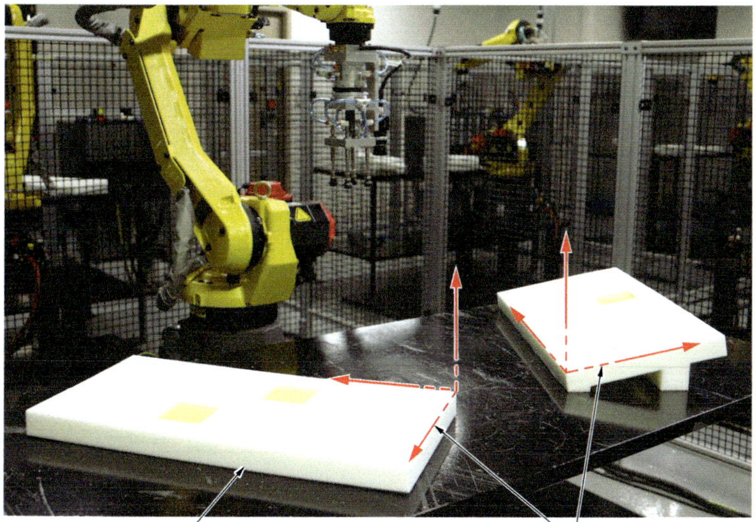

Figure 3-10. A user frame is created for other objects within a cell that the robot must interact with.

Palletizing Software Option

A common robot application in industry is for packaging and stacking items for shipping. If a robot must stack boxes on a pallet, for example, a user frame is created for the position of the empty pallet. The robot then references that user frame as it stacks boxes in the desired pattern for space efficiency and stability.

This process can be programmed in the usual manner. But since palletizing is a common robot task, FANUC developed a software option called Palletizing that simplifies the process of creating palletized layers of product. This option can save hours of programming time. After the first layer of product is created, it enables additional/alternate layers to be added and rotated, resulting in a stable, loaded pallet. This and other software options can be purchased and added at any time but are less expensive if purchased at the same time as the robot.

A user frame is created on a certain fixture, and all points interacting with this fixture reference that user frame. If the fixture changes position or is replaced, the programmer must re-create the user frame. If a user frame is not created, all points would have to be based off the origin of the World frame, which is less practical and intuitive.

There are three methods for creating a user frame: three-point, four-point, and direct entry. The three-point method of creating a user frame is the most commonly used method.

Using the Three-Point Method to Establish User Frames

In the three-point method, the robot is first taught that a certain point in space is the origin of the new user frame. The origin point is usually the corner of the object or fixture. **See Figure 3-11.** The robot must be jogged precisely to the selected point with the end effector's TCP gently touching it. To create the user frame, the robot must be jogged in either the World or joint method. Also, an accurate tool frame must be selected in the Jog menu prior to creating the first point of the user frame. **See the *Establishing a User Frame Using the Three-Point Method* Procedure.**

Three-Point Method for User Frames

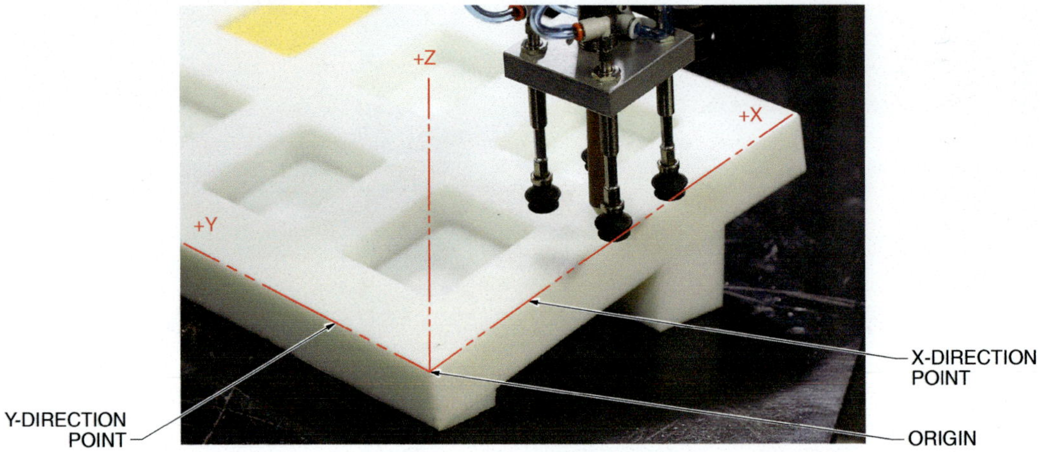

Figure 3-11. The three-point method uses an origin point plus X-direction and Y-direction points to define a user frame.

Next, the robot is taught the orientation of the new user frame by assigning X and Y directions from the origin point. This is done by moving the TCP some distance in the desired X direction, recording that point, and then repeating the procedure for the Y direction. The three points must form a 90° angle, and the right-hand rule is useful to ensure that the user frame's +Z direction goes up and away from the fixture.

These X and Y points may be the corners of a fixture, any point along the side of the fixture, or some other point in the user-defined X and Y directions. They must be at least 100 mm (4″) from the origin point, but distances farther than that provide greater accuracy in the user frame definition. Of course, all three points must be within reach of the robot. A common aid for quickly and accurately creating user frames is to have permanent reference marks for the three points on the fixture, such as small divots from the tip of a drill bit.

PROCEDURE

Establishing a User Frame Using the Three-Point Method

Choose the User Frame to Set Up

1. On the teach pendant, press the MENU button.
2. Use the arrow buttons to select `SETUP`.
3. Use the arrow buttons to move to the right, select `Frames`, and press ENTER.
4. Verify that the top-left of the display shows `User Frame`. If it does not, then change the frame type:
   a. Press F3 `[OTHER]` to show the `OTHER` menu.
   b. Use the arrow buttons to select `User Frame` and press ENTER.

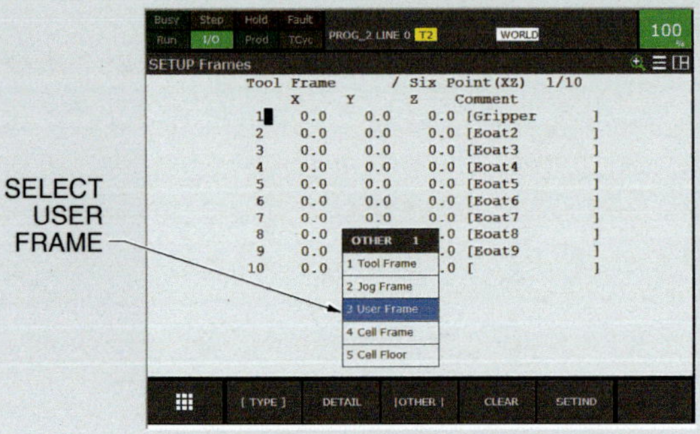

⑤ Use the arrow buttons to select the desired user frame to establish.

Choose Method

⑥ Press F2 `DETAIL` to enter the `SETUP Frames` screen.

⑦ Press F2 `[METHOD]` to show the `METHOD` menu.

⑧ Use the arrow buttons to select `Three Point` and press ENTER.

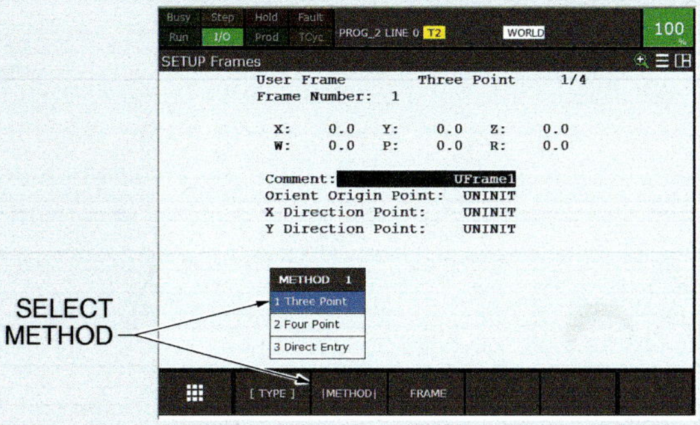

Name the User Frame

⑨ Use the arrow buttons to select the `Comment` entry and press ENTER.

⑩ Use the arrow buttons to select `Options/Keybd`.

⑪ Press F5 `KEYBOARD`.

⑫ Using the teach pendant keyboard, enter a comment or name for the user frame and press EXIT. (Remember that no spaces or unique characters can be used.)

Record the Origin Point

⑬ Use the arrow buttons to select `Orient Origin Point`.

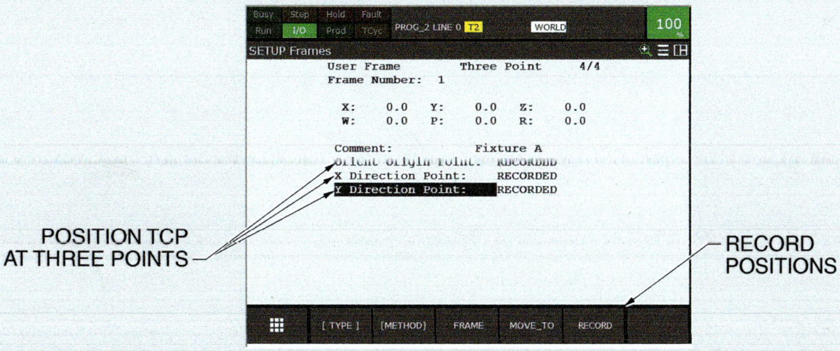

⑭ Jog the robot to the origin point for the object the user frame is being created on.

⑮ Press and hold SHIFT and then F5 `RECORD`.

Record the User Frame Orientation

⑯ Use the arrow buttons to select `X Direction Point`.

⑰ Jog the robot along the fixture in the +X direction to the desired point.

⑱ Press and hold SHIFT and then F5 `RECORD`.

⑲ Use the arrow buttons to select `Y Direction Point`.

⑳ Jog the robot along the fixture in the +Y direction to the desired point.

㉑ Press and hold SHIFT and then F5 `RECORD`.

㉒ Press PREV to return to the `SETUP Frames` screen.

Using the Four-Point Method to Establish User Frames

The four-point method is completed with the same procedure as the three-point method but with the addition of a fourth point, which is called the system origin. This point is the exact center of the fixture. The four-point method is only required when the robot is integrated with a camera. This integration process is called Vision by FANUC.

Using the Direct-Entry Method to Establish User Frames

To create a user frame using the direct-entry method, an accurate and precise drawing of the robot cell must be created. All measurements shown on the drawing must be referenced from the origin of the robot's World frame. After the robot is placed, all equipment to have a direct-entry user frame must be precisely placed to the dimensions in the drawing. Placing large equipment accurately is very difficult. For this reason, direct entry of a user frame is rarely used. However, if done correctly, it can save the programmer a significant amount of time. **See the *Establishing a User Frame Using the Direct-Entry Method* Procedure.**

PROCEDURE

Establishing a User Frame Using the Direct-Entry Method

Choose the User Frame to Set Up

1. On the teach pendant, press the MENU button.
2. Use the arrow buttons to select `SETUP`.
3. Use the arrow buttons to move to the right, select `Frames`, and press ENTER.
4. Verify that the top left of the display shows `User Frame`. If it does not, then change the frame type:
   a. Press F3 `[OTHER]` to show the `OTHER` menu.
   b. Use the arrow buttons to select `User Frame` and press ENTER.
5. Use the arrow buttons to select the desired user frame to establish.

Choose Method

6. Press F2 `DETAIL` to enter the `SETUP Frames` screen.
7. Press F2 `[METHOD]` to show the `METHOD` menu.
8. Use the arrow buttons to select `Direct Entry` and press ENTER.

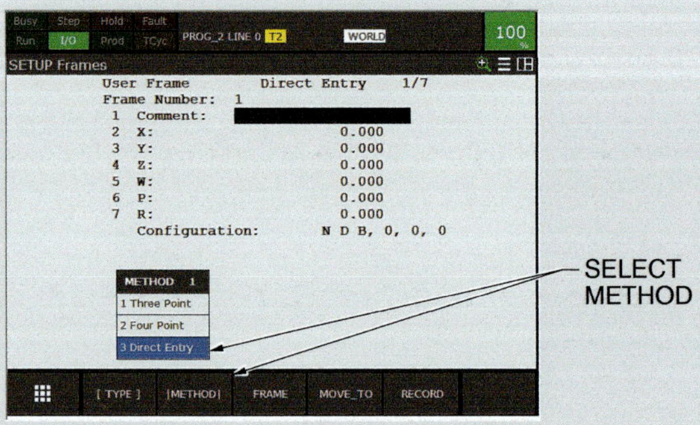

Name the User Frame

⑨ Use the arrow buttons to select the `Comment` entry and press [ENTER].

⑩ Use the arrow buttons to select `Options/Keybd`.

⑪ Press [F5] `KEYBOARD`.

⑫ Using the teach pendant keyboard, enter a comment or name for the user frame and press [EXIT]. (Remember that no spaces or unique characters can be used.)

Enter Dimensions

⑬ Use the arrow buttons to select the first direction or orientation offset line and press [ENTER].

⑭ Enter the associated dimension (in millimeters) from the schematic and press [ENTER].

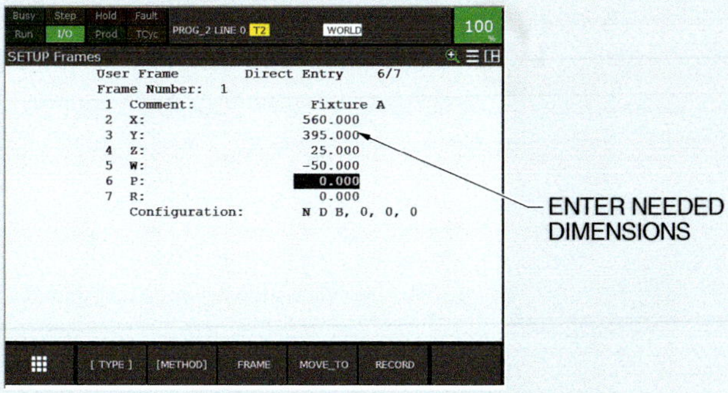

⑮ Repeat steps 13 and 14 for each direction or orientation offset line (as needed).

⑯ Press [PREV] to return to the `SETUP Frames` screen.

Testing User Frames

A user frame should be tested to ensure that it is properly aligned with the fixture or object. A user frame is tested by positioning the robot's TCP slightly above the fixture or object and then jogging the robot in the X and Y directions. **See the *Testing a User Frame* Procedure.**

PROCEDURE

Testing a User Frame

① Jog the robot so that the end effector's TCP is slightly above the user frame's origin point.

② On the teach pendant, press and hold the [SHIFT] button and then press [COORD] to show the Jog menu.

③ Use the arrow buttons to select `User`.

④ Press the number of the user frame to be tested. The Jog menu then automatically disappears.

⑤ Press [COORD] until the jog method of `User` appears.

⑥ Jog the robot in the X, Y, and Z directions.

If the user frame was created properly, the end effector accurately follows the +X and +Y directions of the fixture, such as along its edges. **See Figure 3-12.** Jogging in +Z should move the robot up and away from the fixture. If when jogging in +Z, the robot moves down toward the fixture, the +X direction and +Y direction have been inverted, and the user frame should be re-created. The right-hand rule is used to ensure that +Z moves up and away from the fixture.

If, when jogging the robot in the +X or +Y direction, the robot does not move in the correct direction, the programmer must verify that the correct user frame has been selected by opening the Jog menu and reselecting the user frame number. Then the user frame can be retested. If the robot still does not follow the fixture edge, the user frame must be re-created.

Testing User Frames

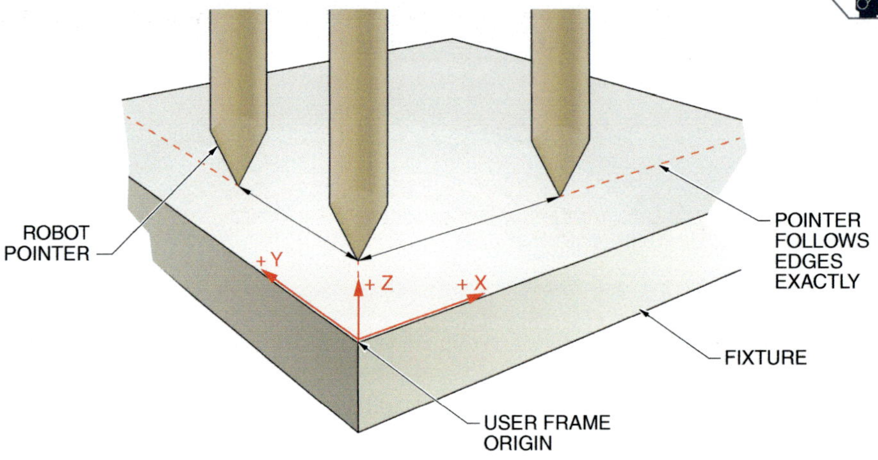

Figure 3-12. When testing a user frame, the robot pointer should follow the edges of the object exactly.

SECTION 3.4 CREATING JOG FRAMES

A *jog frame* is a robot frame with a user-defined origin and X, Y, and Z-axis directions and is used to orient the robot through the space of a robot cell. The purpose of a jog frame is to assist the programmer with maneuvering the robot around the cell. The robot can be jogged while in a user frame around the cell where fixtures are placed and user frames have been created. However, if a user frame was not required on an item that the programmer must navigate around, it may be to the programmer's advantage to create a jog frame for that item.

There are two methods for creating a jog frame: three-point and direct entry. Both methods are used in the same way as when creating a three-point user frame and a direct-entry user frame. The only difference is the selection of a jog frame in the Jog menu. Selecting and testing a jog frame is also the same as for user frames. The COORD button is pressed until `JGFRM` is displayed. The robot then uses the newly created jog frame to move.

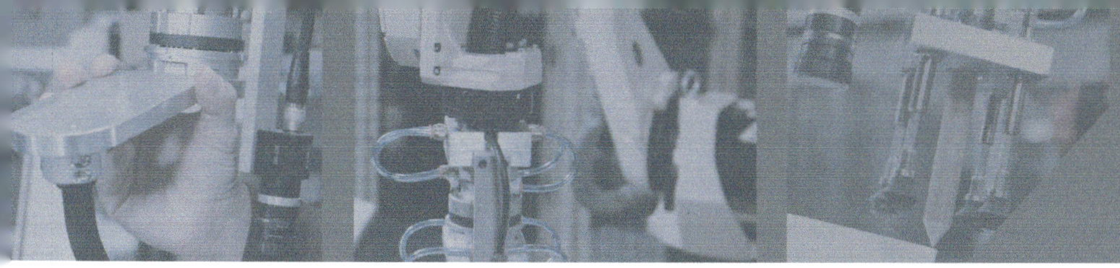

Review 3

Name ______________________________ Date ______________

SECTION 3.1 REFERENCING FRAMES

______________ **1.** Selecting a jogging method/frame is only for ___.
- A. adding position points
- B. moving the robot into desired positions
- C. setting up other frames
- D. running programs

2. How is a frame selected when programming a robot, such as when adding position points?

3. What is the only frame created by the robot manufacturer?

SECTION 3.2 CREATING TOOL FRAMES

______________ **4.** A tool frame is a robot frame with an origin at the ___.

______________ **5.** A robot can have up to ___ different tool frames.
- A. 3
- B. 10
- C. 64
- D. 100

______________ **6.** Each tool frame is used for a different ___.

______________ **7.** The direct-entry method of creating a tool frame requires a ___.
- A. fixture
- B. fixed pointer
- C. schematic of the end effector
- D. schematic of the robot cell

______________ **8.** Creating a three-point tool frame requires ___ pointer(s).
- A. 1
- B. 2
- C. 3
- D. 4

_____________________ **9.** In the three-point method of creating a tool frame, the ___ pointer is located anywhere within the robot cell.

_____________________ **10.** The six-point (XZ) method must be used for certain end effectors, such as a ___.

A. welding torch
B. pointer
C. gripper
D. vacuum picker

_____________________ **11.** After teaching a robot the location of the TCP with the approach points, the ___-point (XZ) method adds extra steps to teach the robot which directions will be considered X and Z.

12. Where is the factory-established TCP?

13. Why must an end effector be new in order to use the direct-entry method of creating a tool frame?

14. How does the three-point method for creating a tool frame allow a robot to define the location of the TCP?

15. If a tool frame has been created properly, what is the result of the tool frame test?

SECTION 3.3 CREATING USER FRAMES

_____________________ **16.** A user frame is used to orient a robot to ___.

A. the perimeter of the cell
B. the operator
C. Axis 1
D. an object within the cell

_____________________ **17.** An accurate ___ frame must be selected in the Jog menu prior to creating the first point of a user frame.

______________________ **18.** In the three-point method of creating a user frame, the robot is first taught that a certain point in space is the ___ of the new user frame.

A. reference point
B. X-direction point
C. origin
D. TCP

______________________ **19.** The three points used in the three-point method of creating a user frame must form a(n) ___° angle.

______________________ **20.** In the four-point method of creating a user frame, the fourth point is the ___ of the fixture.

A. center
B. opposite corner
C. origin
D. Z-direction point

______________________ **21.** When using the direct-entry method of creating a user frame, all measurements must be referenced from the origin of the ___ frame.

22. Where is the origin of a user frame?

23. What does the direct-entry method of creating a user frame require?

24. How is a user frame tested?

SECTION 3.4 CREATING JOG FRAMES

25. What is the purpose of a jog frame?

26. What are the two methods for creating a jog frame?

3
Lab Exercise

Lab Objective

Create both a tool frame and a user frame.

Equipment Required

- Robot with a gripper-style or welding torch end effector
- Pointing tool, such as a pen, marker, or pointed dowel
- Fixed pointer, attached to a base or table in front of the robot
- Square object, fixed or taped to the base in front of the robot
  Note: The object should be angled with respect to the World frame. To add more complexity, the object can also be tilted.

Lab Procedure

1. With the pointing tool securely attached to the robot, create a tool frame using the three-point method (or six-point method if using a welding torch.)
2. Test the newly created tool frame. If it was created accurately, the tip of the pointing tool should not move more than a few millimeters from the fixed pointer when Axes 4 and 5 are jogged while the robot is in the World jogging method.
3. Create a user frame on the object using the newly created tool frame. Be sure to select the proper tool frame in the Jog menu.
4. When completed, change the jog method to User and enter the number of the user frame in the Jog menu. The robot should now follow the edges and plane of the object when the major X and Y-axes are pressed. When +Z is pressed, the robot should move up and away from the object.

4
Starting Programs

New programmers must learn the differences between types of programming files, the most common of which is a TP program. There are minimum requirements for setting up the shell of a new TP program. Payload schedules must be developed to account for both the mass and the center of mass of the end effectors and the objects they handle. Then movement lines can be added into the program. These lines of code are the fundamental parts of any robot program that move the robot between points in space. Basic editing commands are used to change the number and order of those program lines.

Objectives

SECTION 4.1 IDENTIFYING FILE TYPES

- Differentiate between the different types of programming files.
- Describe when to use a macro program.
- Describe the purpose of a system variable file.

SECTION 4.2 SETTING UP TP PROGRAMS

- Demonstrate how to create a new program shell, name it, and add lines.
- Describe the most commonly used parameters in the Program Detail screen.

SECTION 4.3 CREATING PAYLOAD SCHEDULES

- Describe the importance of an accurate payload schedule.
- Explain when to create multiple payload schedules.

SECTION 4.4 USING EDITING COMMANDS

- Describe how to access the editing commands.
- Describe how to use each of the commonly used commands.

SECTION 4.5 PROGRAMMING MOVEMENT LINES

- Explain the importance of establishing a Home position.
- Demonstrate the procedure for creating a movement line.
- Describe each part of a movement line.
- Describe how to re-create an existing point.
- Demonstrate how to use the Touchup command.

Key Terms

- teach pendant (TP) program
- macro (MR) program
- KAREL program
- KAREL
- system variable (SV) file
- program shell
- group
- group mask
- payload
- payload schedule
- comment
- movement line
- Home position
- movement type
- position ID
- termination type

Learner Resources

atplearningresources.com
Access Code: **762316**

SECTION 4.1 IDENTIFYING FILE TYPES

There are three types of programming files used when creating FANUC robot programs. **See Figure 4-1.** The primary programming file used is a teach pendant program. A teach pendant program file can be used to complete nearly any application. This is the main file type for creating robot programs. The second programming file type is called a macro. Only small, reoccurring sets of program lines are used in a macro. The small macro programs are then called into teach pendant program files. The third type of programming file is called a KAREL program. KAREL programming is only required in a few select applications.

A system variable file is a fourth type of file used with a FANUC robot, but this type is not a programming file. This file is established by FANUC to enable a programmer to set the values of certain operating characteristics, which changes how the robot reacts in different situations.

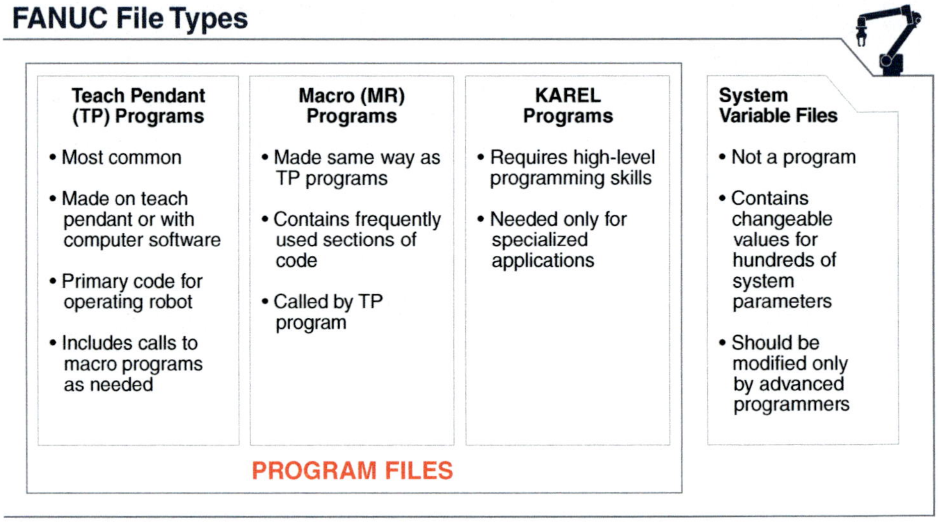

Figure 4-1. Files used in FANUC robots include three types of program files and system variable files.

Teach Pendant Programs

A *teach pendant (TP) program* is a computer file containing lines of code for controlling robot actions. Writing TP programs is the most commonly used method of programming a robot. TP programs can be created directly on the teach pendant, or they can be developed in FANUC's ROBOGUIDE software for Windows® computers. **See Figure 4-2.** In the software, programs are developed in a virtual teach pendant, so the programming procedures are identical. Programs created on a computer are then transferred to the controller for execution.

TP programs are made of two types of program lines: movement lines and command lines. In the following example of the Home Program, the first two lines are command lines and do not result in robot movement by themselves. The third line is a movement line that instructs the robot to return to the Home position.

```
:   !Home Program
:   UTOOL_NUM=1
:   UFRAME_NUM=0
:J   P[1:Home] 100% FINE
```

Teach Pendant (TP) Programs

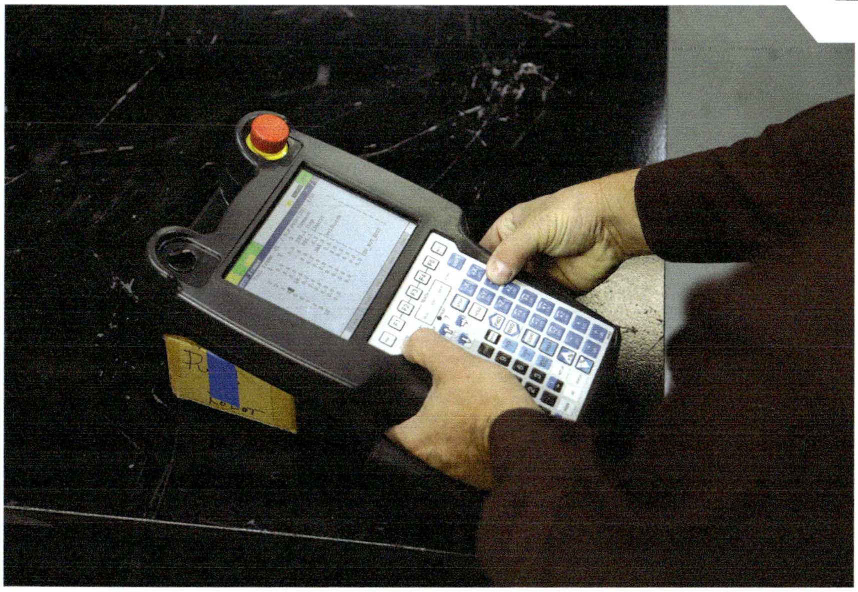

PROGRAMMING ON TEACH PENDANT

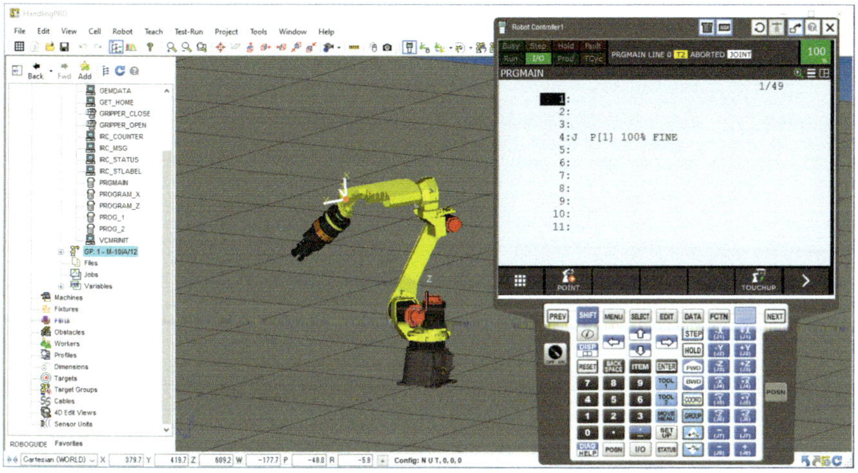

PROGRAMMING IN COMPUTER SOFTWARE

Figure 4-2. TP programs can be created on a teach pendant or with computer software.

Macro Programs

A *macro (MR) program* is a computer file containing a small section of code for a particular, commonly repeated action. The code is written in the same manner as a TP program. MR programs are limited on the type of commands that can be used. However, they are an excellent means of grouping routine sections of code that are frequently used in a robot program. Programmers also refer to these reoccurring sections of code as subroutines.

For example, an MR program may be created with the code to open an end effector or to move a robot back to its Home position. Then, instead of including the code multiple times within a TP program, the name of the MR program can be called in only one line. **See Figure 4-3.** Each time the software reaches that MR program name, it executes that macro code and then returns to the TP program. While not strictly required in order to program the robot, this is a common practice to maintain efficient and readable code. Also, if changes are needed to the macro code, those changes are needed in only one location to affect all instances of that code.

In the example Home Program, a "Home" macro program is used to send a robot back to its Home position. Care should be taken when running programs such as this. The robot always takes the shortest path back to the position designated, which may interfere with objects in its path and cause a crash.

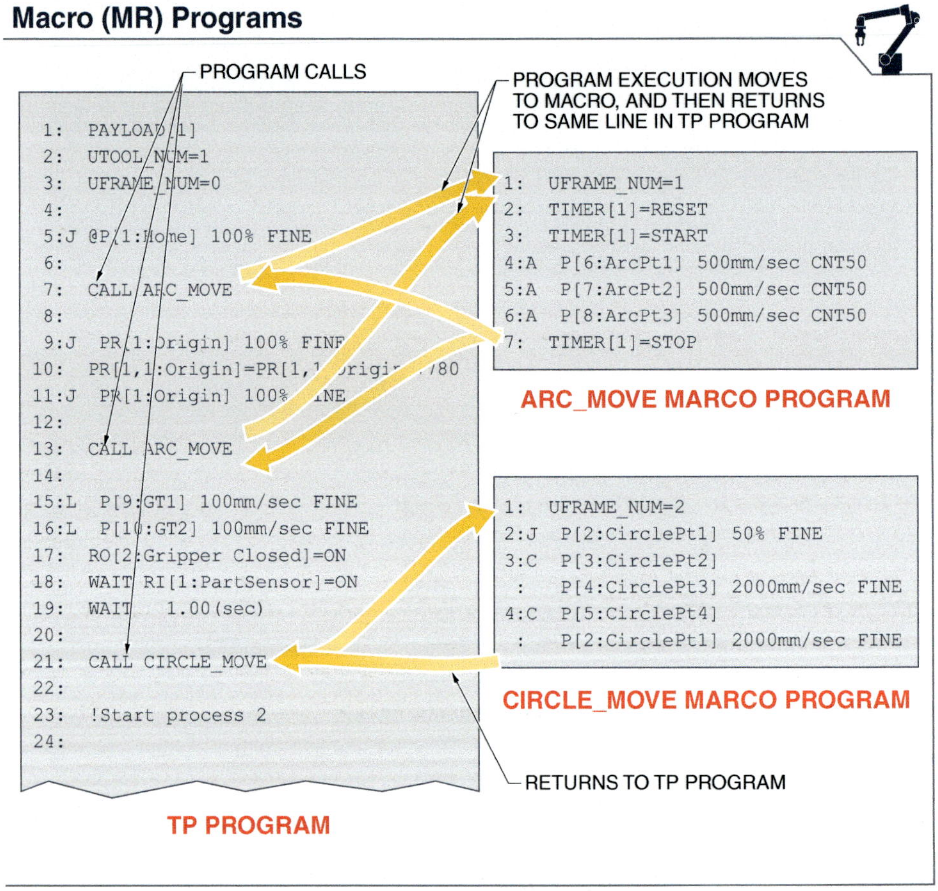

Figure 4-3. MR programs are called from TP programs whenever they are needed.

KAREL Programs

A *KAREL program* is a computer file containing lines of code written in KAREL. *KAREL* is a Pascal-based programming language designed specifically for controlling FANUC robots. **See Figure 4-4.** KAREL provides a precise and specific level of control that TP programming is not capable of, but it is needed in only a few select situations. It also requires that the robot be compatible with KAREL programs. Robot manufacturers have moved away from programming in such languages in order to make the process easier for the everyday user. Today, almost any FANUC robot can be programmed using only TP programs.

KAREL Programs

```
IF (status = cc_success) THEN
  SELECT pos_type OF
    CASE (cc_xyzwpr):
      xyz= GET_POS_REG(1, status)
    CASE (cc_jntpos):
      jpos = GET_JPOS_REG(1, status)
      xyz = jpos
    ELSE:
      write ('The position register set to invalid type', pos_type,CR)
      status = -1
  ENDSELECT
  IF (status = cc_success) THEN
    xyz.x = xyz.x+10
    xyz.z = xyz.z-10
    jpos  = xyz
    WRITE('Setting New Position to PR[2]',cr)
    SET_JPOS_REG(2,jpos,status)
    IF (status = cc_success) THEN
      WRITE('Setting Integer Value to R[1]',cr)
      SET_INT_REG(1, 10, status)
```

Figure 4-4. KAREL is a high-level programming language for controlling FANUC robots.

System Variable Files

A *system variable (SV) file* is a computer file containing a list of parameters and their values, which can be adjusted to change the way the teach pendant and robot function. **See Figure 4-5.** Changes to these files can impact anything from the way the robot reacts when jogging to how programs react when running. For example, a programmer can set a system variable that limits the speed to 10% of full speed when jogging the robot. There are several hundred different system variables that can be adjusted.

Extreme caution should be taken while accessing SV files. Poor choices in the changes, even with warning to the operator, could cause serious operator injury or robot damage.

System Variable (SV) Files

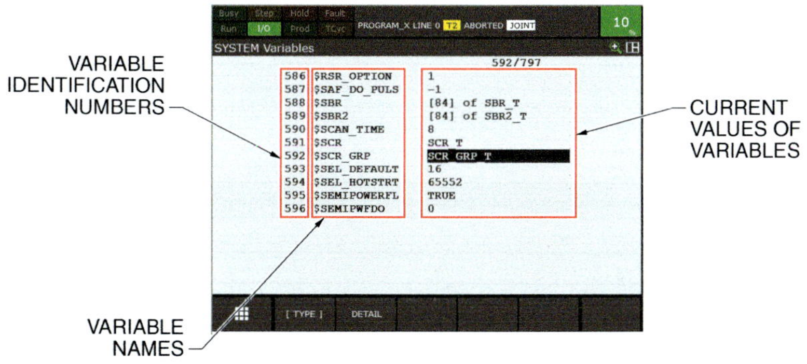

Figure 4-5. Hundreds of different parameters can be adjusted manually in the SV file.

SECTION 4.2 SETTING UP TP PROGRAMS

Both the tool frame and all user frames must be created prior to creating any program. Otherwise, the program positions will be inaccurate. A program shell is then created.

Program Shells

A *program shell* is a program file containing the program's name and empty lines but does not yet contain any code. The programmer enters the robot's movements and other commands into the program shell. Program lines are numbered consecutively, beginning with line 1, and a program shell can contain up to 32,766 lines. **See the *Creating a Program Shell* Procedure.**

PROCEDURE

Creating a Program Shell

① On the teach pendant, press the SELECT button to show the `Program Selection` screen.

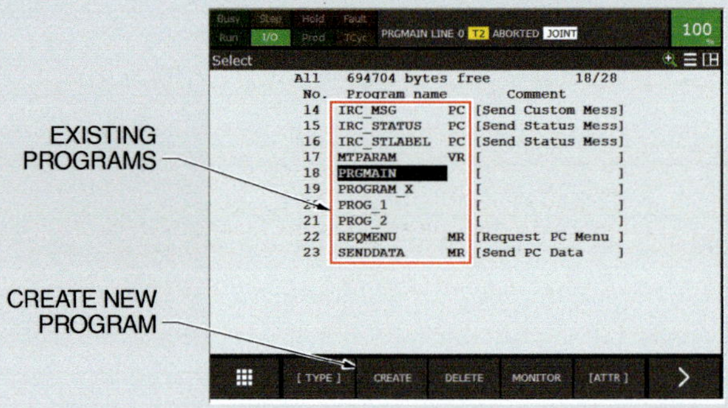

② Press F2 `CREATE` to start a new program. (If `CREATE` is not shown, press NEXT. The > symbol in the lower-right corner of the screen indicates additional function selections available.)

③ A menu appears in the lower-right area of the display. To name the program, use the down arrow button to highlight `Options/Keybd` and press ENTER.

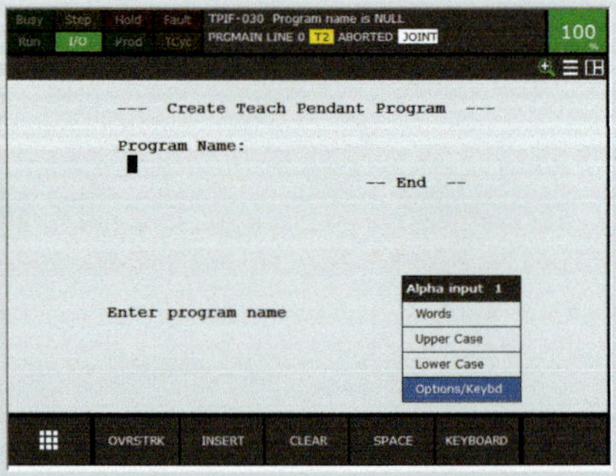

④ Press F5 `KEYBOARD`.

⑤ Type in the desired program name and press EXIT. (A program name may have no more than 32 characters. Also, there may be no spaces or special characters other than underscores.)

⑥ Press F3 `EDIT` to show the program-editing screen.

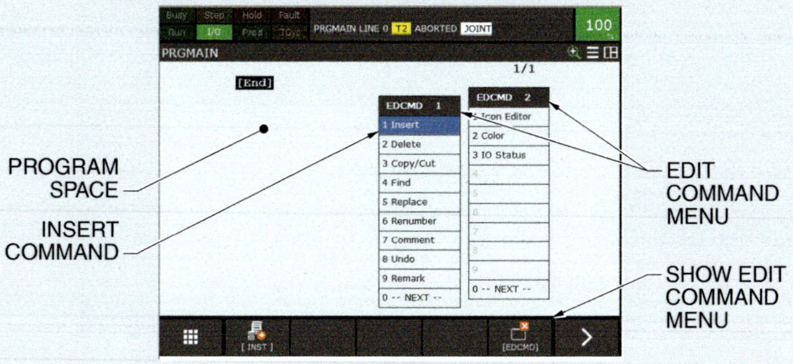

⑦ Press F5 `[EDCMD]` to show the `Edit Command` menu. (If `[EDCMD]` is not shown, press NEXT first.)

⑧ Select the `Insert` command.

Note: Depending on software version and menu settings, the `EDCMD` menu may have a different appearance, and the `Insert` command may already be highlighted. Press ENTER.

⑨ The screen displays: `How many line to insert?` Type `50` with the number pad and press ENTER.

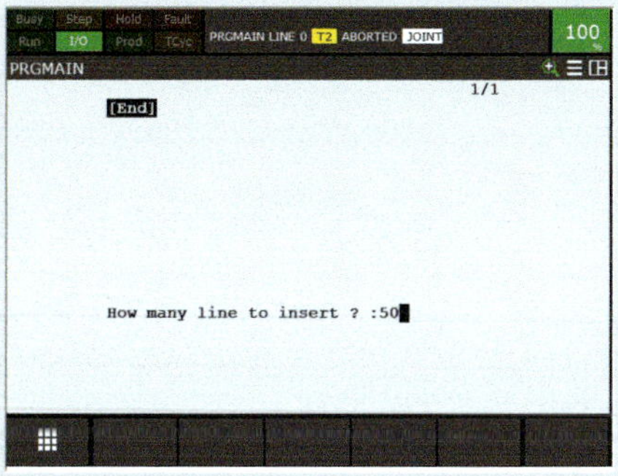

Lines numbered 1 through 50 will then appear. **See Figure 4-6.** After the 50th line, the word `[END]` appears. If more lines are needed later, they can be added at any time. There is never a problem with having too many lines. Lines can also be deleted or left blank in the program to separate tasks. Blank lines do not take up processor time or create issues.

Program Details

At this point, the program is initialized and ready for the programmer to begin creating code. Before entering code, the programmer must review the program details. This ensures the program is set up as desired. **See the *Accessing Program Details* Procedure.**

Program-Editing Screen

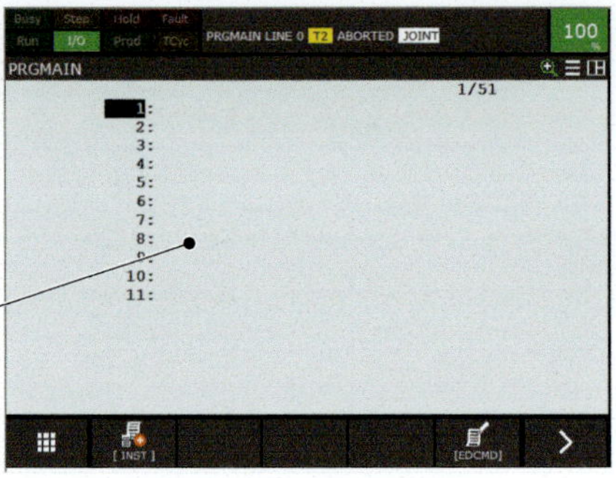

Figure 4-6. There is no problem with having many empty lines in the program space. They do not require processing time and can be used to divide parts of the program with white space.

PROCEDURE

Accessing Program Details

① On the teach pendant, press the SELECT button to bring up the `Program Selection` screen.

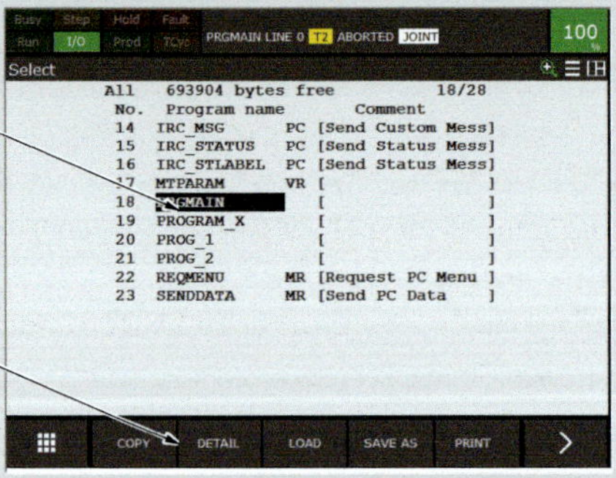

② Use the arrow buttons to highlight the desired program name. The most recently saved program may already be highlighted.

③ Press F2 `DETAIL` to enter the `Program detail` screen. (If `DETAIL` is not shown, press NEXT first to view additional functions.)

The Program Detail screen displays basic information about the program. **See Figure 4-7.** It also includes several parameters that can be modified. The up and down arrows are used to move through the list. Some function key options change depending on the highlighted parameter. For now, it is not necessary to understand every parameter and option, but it is important to understand the most commonly used ones.

Program Name. The first line shows the name of the program. From this screen, the program name can be changed. Names are limited to 32 characters, and no spaces or special characters (other than underscores) may be used.

Program Detail Screen

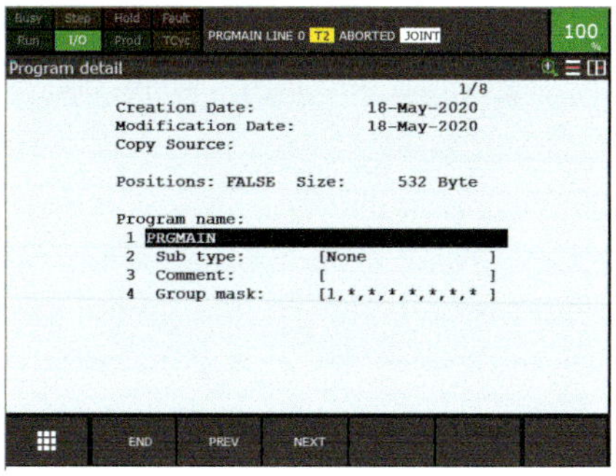

Figure 4-7. The Program Detail screen shows information and changeable parameters for a program.

Sub Type. The sub type parameter defines the type of program file. Highlighting this space provides the programmer with the function F4 `[CHOICE]`. From this menu, the program can be left as `None`, which indicates a TP program, or changed to an MR program.

Comment. Additional comments can be placed in this space. A programmer might provide a very brief description of what the program does or if it is a subprogram to a larger, main program.

Group Mask. A *group* is a set of axes, such as the robot's axes, and also possibly other electromechanical devices that could be controlled by the robot program. Examples include turntables, conveyors, or linear actuators. The *group mask* is a list of possible axis groups with indications of which are active and available. **See Figure 4-8.** The first position signifies the robot. A number "1" in this position indicates that the robot axes are active. Any additional axes connected to the controller would be notated by a "1" instead of the asterisks (*) in positions 2 through 7. If the robot only shows the first position active, the robot is not equipped with additional axes.

Group Masks

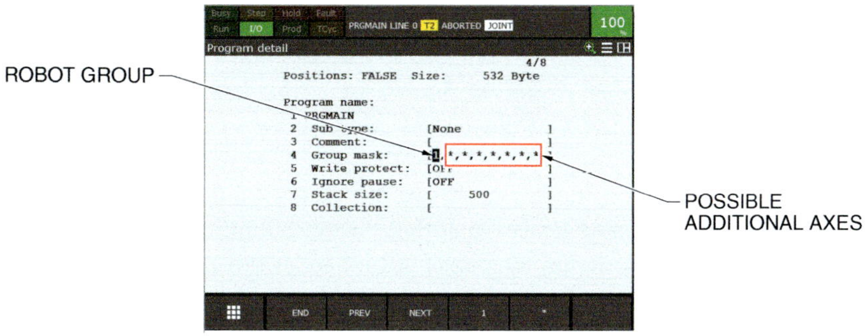

Figure 4-8. The group mask is used to disable controls for extra axes that do not exist or are not being used for a particular program.

Some programs may not use additional axes, even if they are available. This space on the Program Detail screen allows the programmer to turn those axes off. If the axes were left on but not used, the programmer would need to account for this in the program, even though the application does not need them. Turning them off simplifies programming. To turn off additional groups, the "1" can be changed to an asterisk (*).

Write Protect. When the write protect parameter is turned off, anyone can easily make changes to the program. Write protection is turned off and on with the function buttons. Once write protection is turned on, programmers need to consciously go into the program details and change `Write protect` to `OFF` in order to make changes to the program. This type of security simply prevents operators from altering or deleting a line of code by mistake.

Ignore Pause. When the ignore pause parameter is turned off, a programmer or operator can pause a program by pressing HOLD. When the parameter is turned on, the HOLD button does not pause the program. This prevents operators from stopping the robot mid-program.

Stack Size. A program can call another program, such as an MR program, as a way to easily insert portions of frequently used code. When the called program is finished being executed, the controller returns to the original program where it left off. However, called programs can themselves call additional programs, and those can call programs, and so on. In complex programs, this can continue for many levels. However, if the continued calling without reaching a program `END` statement and jumping back into the main program reaches a set limit, the robot will show a `Stack Overflow` fault. The stack size parameter sets this limit (the number of program calls permitted without reading an `END` statement). Generally, a stack size of 500 is large enough to meet most programs' needs.

SECTION 4.3 CREATING PAYLOAD SCHEDULES

A *payload* is the combined mass of an end effector and any object that is picked up and handled by the end effector. **See Figure 4-9.** A robot may handle many different objects in the course of a program, and information about each object can be put in a payload schedule. A *payload schedule* is a list of payload masses and their associated information.

When the robot knows the mass of a payload, the controller adjusts the robot's speed, giving the servomotors (which run the robot's axes) the ability to stop the robot in the correct position. For lighter payloads, the robot can move faster. When the payload is heavier, the robot must move slower.

In most educational settings, a payload schedule is not required because the total mass and size of the end effector and load are usually not large enough to affect speed and position accuracy. However, if there were a situation in which the payload value is set significantly lighter than the actual load, the robot position could be affected, or the robot could fault with a servomotor overcurrent fault, which is critical for efficient operation. The proper payload schedule ensures a robot's accuracy and appropriate top speed, which is critical for efficient operation.

Payloads

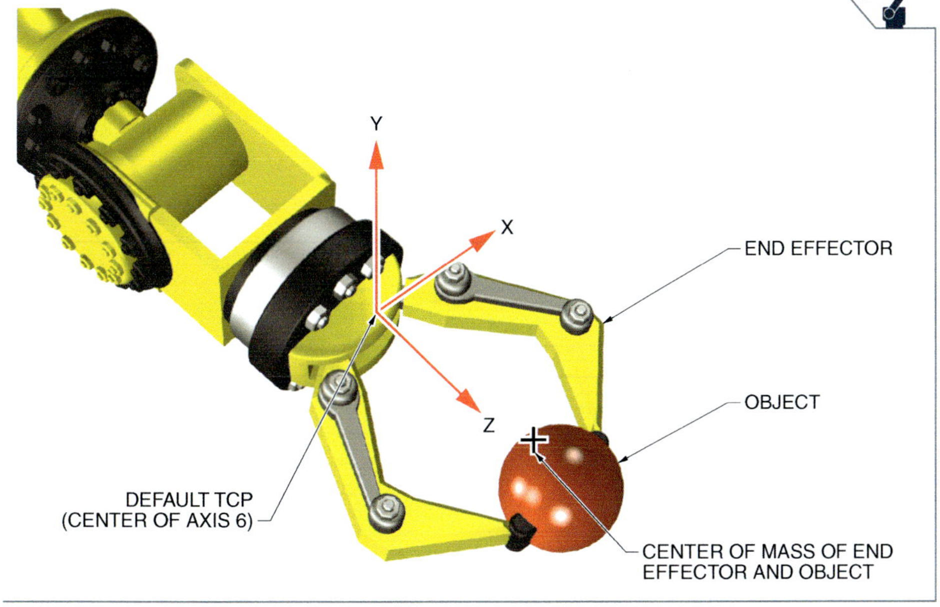

Figure 4-9. A payload is the combined mass of the end effector and any object that it handles. The robot must know the mass, center of mass location, and inertia in order to handle the payload properly.

Controller Fuses

There are three different types of fuses used in a robot controller. New robots come with one extra fuse of each type. However, it is not uncommon to blow fuses, especially in an educational setting, so it is advisable to keep additional fuses of each type on hand.

Robot controller fuses are not the types that can be purchased from the local hardware store, and ordering them takes some time before arrival. Talking with a robot parts supplier or contacting FANUC directly is the best way to purchase the correct fuses and ensure they are on-hand when the need arises.

BACK OF CONTROLLER

Payload information includes the payload's center of mass and moment of inertia. These dimensions are referenced from the center of Axis 6, the robot's default TCP. If this information is critical for the program application, such as for a heavy or unbalanced payload, it will likely be provided by an engineer. With light payloads, it may be adequate to estimate the center of mass by placing the product in the end effector and measuring in the X, Y, and Z directions to the center point of the total mass. It is important to remember that this may not be in the physical center of the payload. **See the *Creating a Payload Schedule* Procedure.**

PROCEDURE

Creating a Payload Schedule

Choose the Payload Schedule to Set Up

1. On the teach pendant, press the MENU button.
2. This menu spans multiple pages. Use the arrow buttons to highlight `--NEXT--` and press ENTER to show the second page.

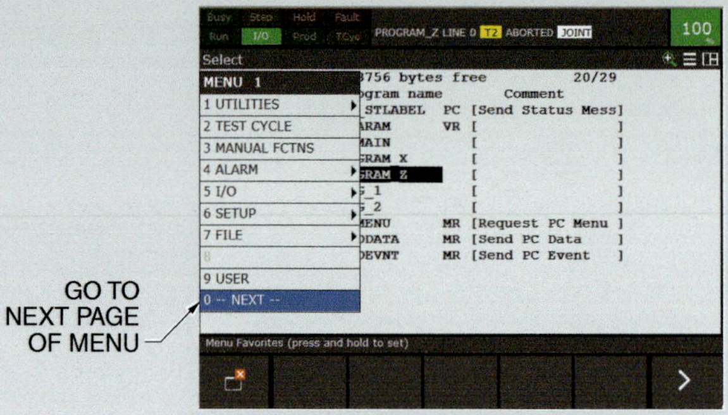

3. Use the arrow buttons to highlight `System`, and a secondary menu appears.

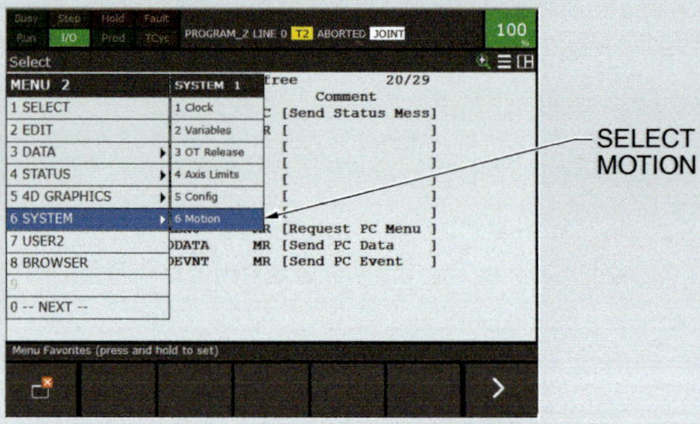

4. Use the arrow buttons to highlight `Motion` and press ENTER to enter the `MOTION PERFORMANCE` (Payload Schedule) screen. Notice that the maximum payload the robot can safely handle is already set.

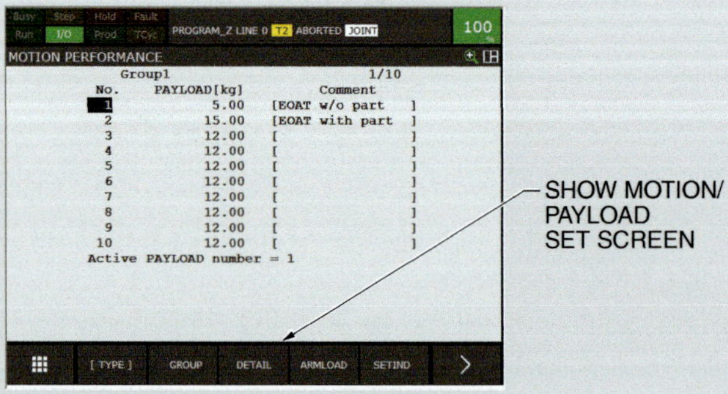

5. Use the arrow buttons to highlight the desired payload schedule.

⑥ Press [F3] `DETAIL` to enter the `MOTION/PAYLOAD SET` screen.

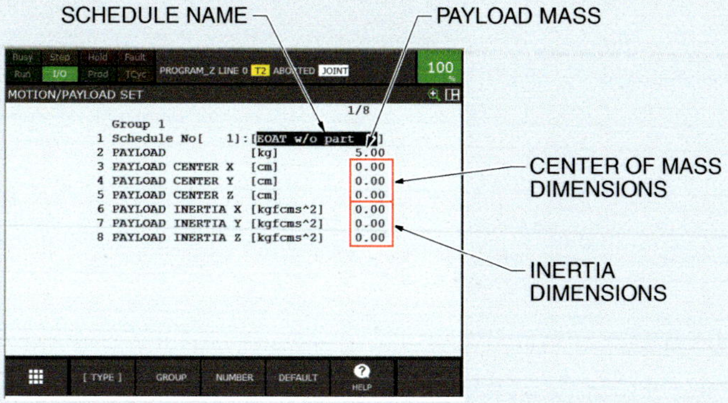

Name the Payload Schedule

⑦ Select the payload name in the first line and press [ENTER].

⑧ Use the teach pendant on-screen keyboard to enter a name for the payload schedule and press [EXIT]. (Remember that no spaces or unique characters can be used.)

Enter Mass and Dimensions

⑨ Use the arrow buttons to highlight `PAYLOAD` (mass) and press [ENTER].

⑩ Enter the mass (in kg) and press [ENTER].

⑪ The screen displays: `Path and Cycletime will change. Set it?` Press [F4] `YES`.

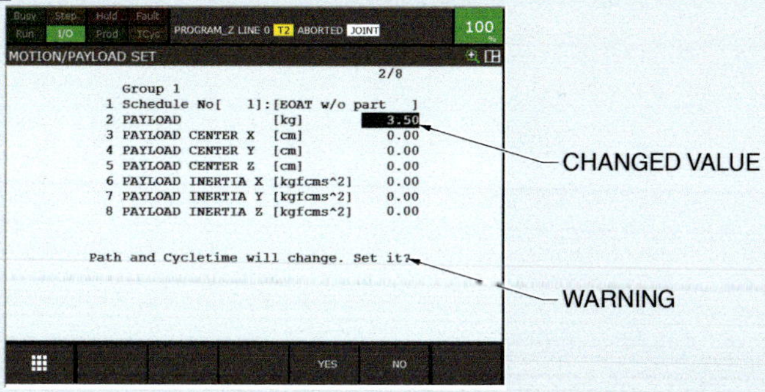

⑫ Use the arrow buttons to highlight each `PAYLOAD CENTER` (of mass) dimension (X, Y, and Z) and press [ENTER].

⑬ Enter the dimensions (in cm) and press [ENTER].

⑭ Press [F4] `YES` to accept the path and cycletime change.

⑮ Use the arrow buttons to highlight each `PAYLOAD INERTIA` value (X, Y, and Z) and press [ENTER].

⑯ Enter the values (in kgf·cm·s^2) and press [ENTER].

⑰ Press [F4] `YES` to accept the path and cycletime change.

⑱ Once the payload is set, press [PREV] to return to the main payload schedule screen.

Payload schedules are created for each tool frame since each tool frame represents a different end effector. There may be multiple payload schedules for each tool frame, accounting for being with and without loads. Each type of load could have a different mass or center of mass. The programmer can make up to 10 different payload schedules. Within a program, payload schedules are then called up as needed as an end effector interacts with different loads. **See the *Adding Payloads to Programs* Procedure.**

PROCEDURE

Adding Payloads to Programs

1. On the teach pendant, press the SELECT button to bring up the `Program Selection` screen.
2. Use the arrow buttons to highlight the desired program name and press ENTER.
3. Within the program, use the arrow buttons to highlight a blank line number.
4. Press F1 `[INST]` to show the `Instruction` menu. (If `[INST]` is not shown, press NEXT first.)
5. Press the right arrow in the menu window to show the second page. Then select `Payload`.
   Note: Depending on software version and menu settings, the `Instruction` menu may have a different appearance.

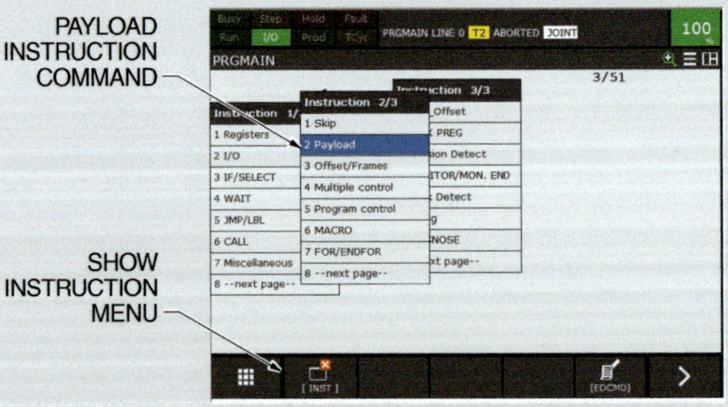

6. Select `PAYLOAD` [...] to insert a program line for calling a particular payload.

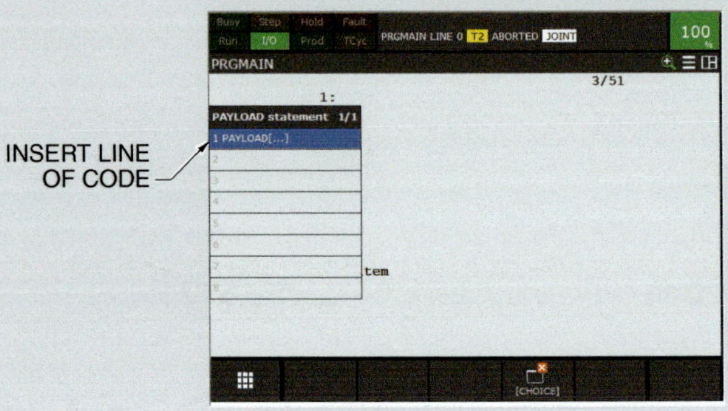

⑦ Enter the desired payload schedule number between the brackets and press ENTER.

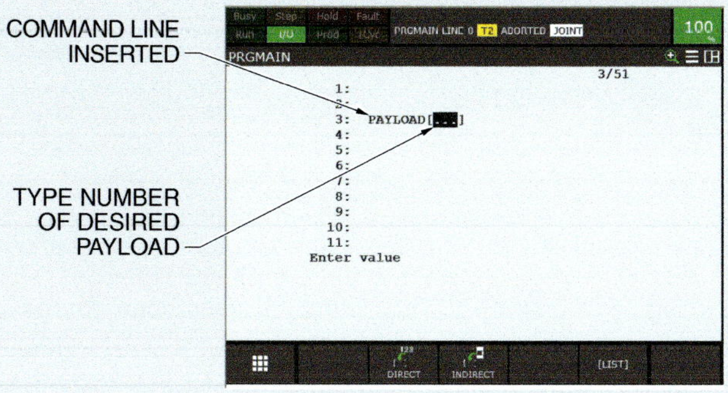

SECTION 4.4 USING EDITING COMMANDS

Once a program shell is created, the `[EDCMD]` (Edit Command) menu is available in the function button area. (It may be necessary to press NEXT to show more functions.) This menu includes a number of commands for changing the number and order of program lines but not the contents of the lines. They are similar to the common edit commands for a text document: cut, copy, paste, and so on.

Pressing F5 `[EDCMD]` will show the menu of program editing commands. **See Figure 4-10.** Several options are available for manipulating program lines. A command can be selected to execute it or to bring up options for that command.

Depending on software version and menu settings, the Edit Command menu may appear as a text list or an icon button view. All common commands are the same, however, regardless of appearance.

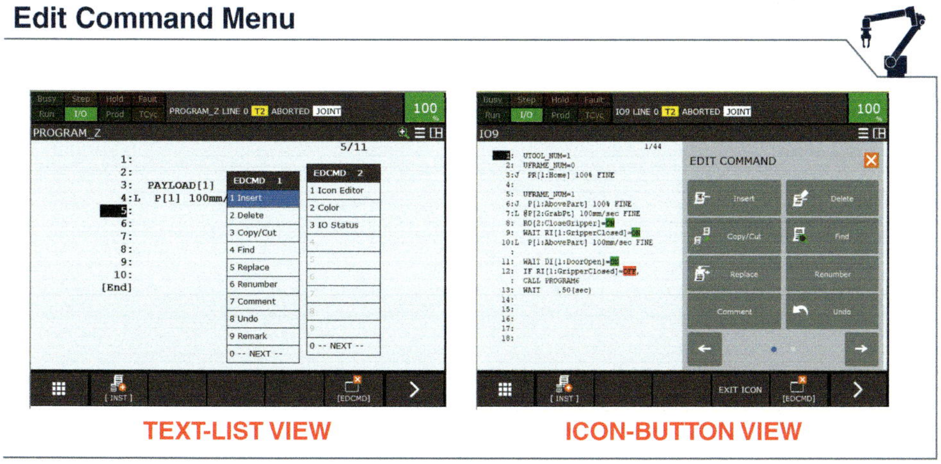

Figure 4-10. The Edit Command menu includes several useful features for ordering and modifying program lines. It may also have different appearances, depending on software settings.

Insert

To insert new, empty lines for code, the arrow buttons are used to highlight the program line number immediately below the point where more lines are desired. Then in the Edit Command menu, select `Insert` and press ENTER. The bottom of the screen displays: `How many line to insert?` Use the teach pendant number pad to enter the number of lines desired and press ENTER. The lines inserted then appear above the highlighted program line.

Delete

Use the arrow buttons to highlight the program line number to be deleted. In the Edit Command menu, select `Delete` and press ENTER. The bottom of the screen displays: `Delete line(s)?` Press F4 `YES`. This deletes only the highlighted line. To delete more than one line at a time, after pressing ENTER, use the arrow buttons to highlight additional lines. When completed, press F4 `YES`.

Notice that F5 `NO` is also an option. This gives the programmer an opportunity to change their mind. If the programmer decides not to delete the lines, pressing F5 `NO` exits the editing menu.

Copy and Paste

Use the arrow buttons to highlight a program line number to be copied. In the Edit Command menu, select `Copy` and press ENTER. Press F2 `SELECT`. If more than one line is to be copied, use the arrow buttons to highlight the range and press F2 `COPY`. Then use the arrow buttons to highlight the program line number where the copied code is to be pasted and press F5 `PASTE`. **See Figure 4-11.** The resulting function buttons provide three paste options:

- F2 `Logic` only pastes the logic. If a movement line is copied, the movement command is pasted, but the position ID is left blank.
- F3 `PosID` pastes the line as shown, including the position ID.
- F4 `Position` pastes a movement line with the coordinates of the robot's position with a new position ID.

When copying and pasting more than one line, pressing NEXT provides five additional options. Each of these options are similar, but the line order is reversed.

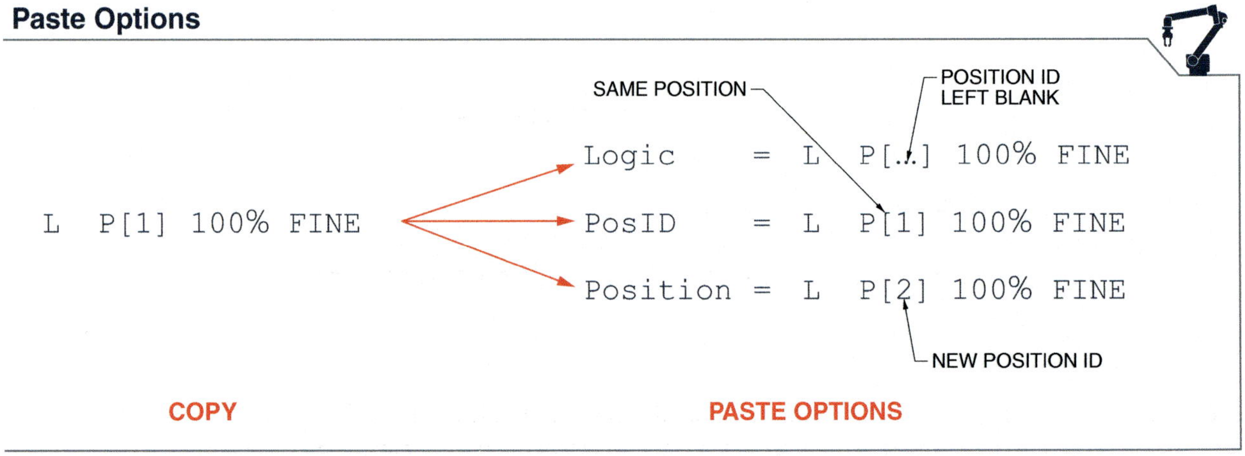

Figure 4-11. When copying movement lines, the Paste function provides three options for handling the position ID.

Comment

The FANUC software allows the programmer to identify registers, inputs, outputs, and other elements with a short string of text. For example, an instruction may be `WAIT DI[1:DoorOpen]=ON`. The "1" specifies the input, but the "DoorOpen" comment makes the input identity clearer—an indication that a door is open. Comments generally make the program easier to understand. However, it may be desirable to make the comments temporarily invisible.

The Comment command toggles the visibility of the comments. In the example, with comments turned off, the instruction line appears as `WAIT DI[1]=ON`. Repeating the command makes the comments reappear. The Comment command does not affect point names.

Undo

The Undo command reverses the last change made. For example, if a line is deleted, the Undo command restores it. If two lines are deleted separately, the command will only restore the last line. The bottom of the screen displays: `Undo? (Delete)`. To complete the command, press F4 `YES`.

Remark

The *Remark command* is a command that designates existing lines that are not to be executed by the software. The software skips over these lines while the program runs. This can be useful when testing or troubleshooting a program to see the effect of leaving certain program lines out of the execution. When using this command, individual lines or ranges of lines can be selected. Then pressing F4 `REMARK` inserts the // characters in front of the program line(s). These characters indicate to the program that those line(s) should be ignored. **See Figure 4-12.** If the command is used again, pressing F5 `UNREMARK` reverses the designation.

This command is different from the Remark instruction. The Remark instruction adds new lines of plain-language text for explanatory purposes.

Remark Command

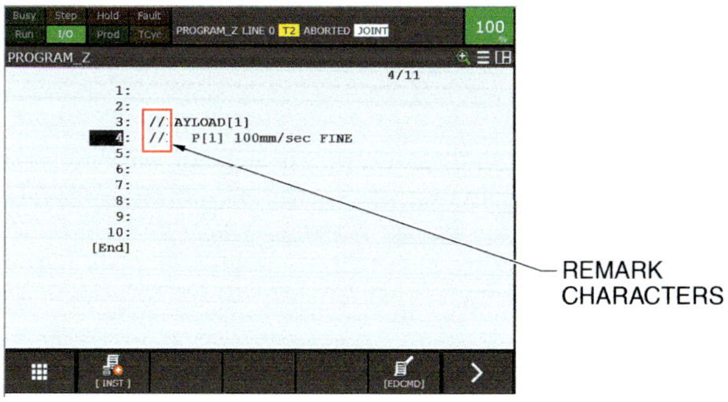

Figure 4-12. The Remark command adds special characters to program lines that instruct the controller to temporarily ignore those lines.

Other Commands

The Edit Command menu includes other commands, such as Find, Replace, and Renumber. These are generally not needed by beginning programmers and may even lead to dangerous robot operating situations if used improperly.

SECTION 4.5 PROGRAMMING MOVEMENT LINES

A *movement line* is a line of code that specifies a particular position in space and the motion specifications to be used when moving to that point. Much of a robot program consists of movement lines as the robot is instructed to move from point to point in its task.

Creating Movement Lines

Most robot programs start from the robot's Home position. The *Home position* is a neutral position for the robot that minimizes the stress on the servomotors and gears. If possible, the Home position also places the end effector where it is easily accessible from the ground. **See Figure 4-13.** This makes maintenance easier on both the robot and the end effector, saving time and reducing the risk of a crash by eliminating the need to jog the robot for access.

Home Position

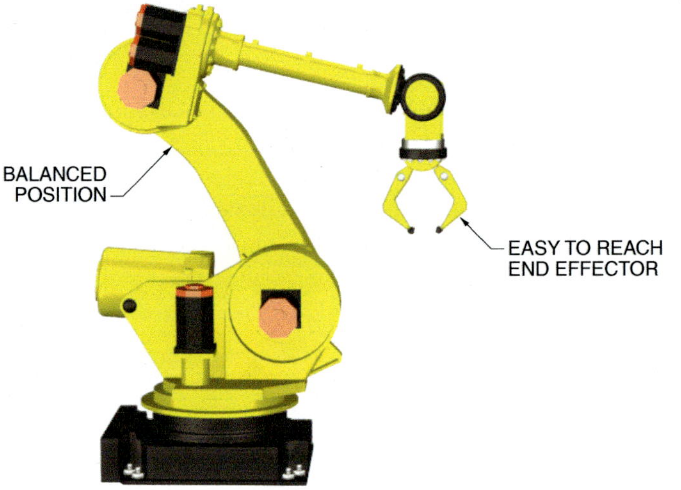

Figure 4-13. The Home position is a safe and convenient position for a robot. It puts the least amount of strain on the joints and motors.

The procedure for creating a movement line applies to any position, but it is best practice that the Home position be the first point and movement line in a program. This position will be used frequently by the programmer and operator. **See the *Adding Movement Lines for Home Position* Procedure.**

PROCEDURE

Adding Movement Lines for Home Position

Select the Program

1. On the teach pendant, press the SELECT button.
2. Use the arrow buttons to highlight the desired program and press ENTER.

Verify the Tool and User Frames

3. Press and hold SHIFT. Then press COORD to bring up the Jog menu.

④ Verify that the required tool frame is identified (by number). If it is not, then change the tool frame:

a. Use the arrow buttons to highlight `Tool`.

b. Use the keypad to enter the number of the desired tool frame. The Jog menu disappears.

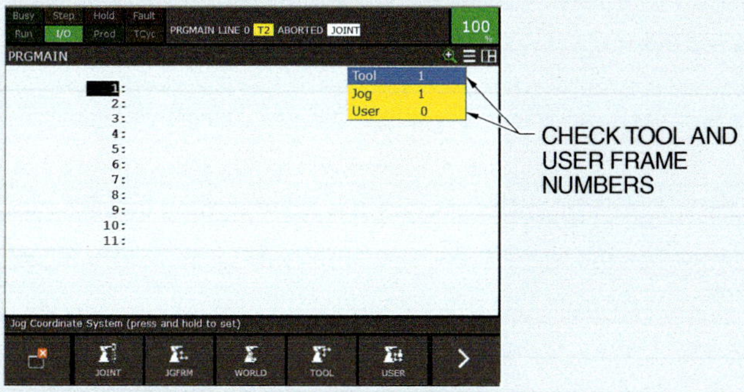

⑤ Press and hold SHIFT. Then press COORD to bring up the Jog menu again.

⑥ Verify that the World user frame (0) is identified. If it is not, then change the user frame:

a. Use the arrow buttons to highlight `User`.

b. Use the keypad to enter 0. The Jog menu disappears.

Insert Movement Line

⑦ Jog the robot to Home position.

⑧ Use the arrow buttons to move to the desired line number.

⑨ Create the movement line for the current position using one of the following:

- Press F1 `POINT`. A menu appears with options for the type of movement and termination desired. Select the desired settings. (If `POINT` is not shown, press NEXT first.)
- Press and hold SHIFT. Then press F1 `POINT` to insert a movement line at 100% speed and Fine termination type.

Note: Once the point is created, the symbol @ appears between the *J* and *P*. This symbol indicates that the robot is currently at that position.

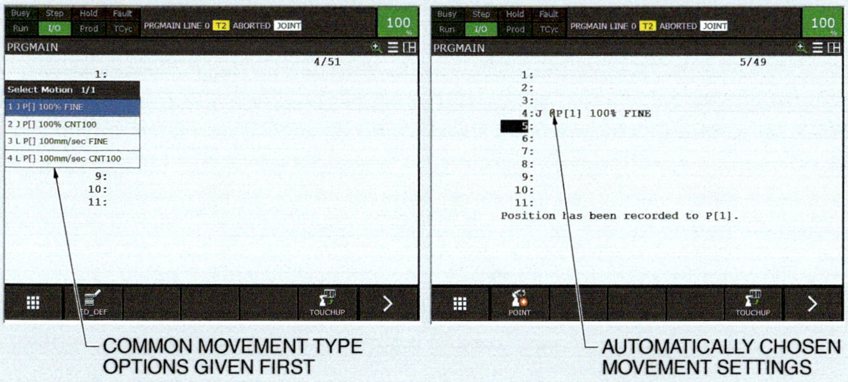

PRESS F1 POINT

PRESS SHIFT AND THEN F1 POINT

The robot can be jogged to different positions, and more movement lines can be added to the program. Verifying the tool and user frames in the Jog menu each time is critical if multiple frames have been created. As a point is created, its position in space is referenced to the currently selected frames. It is common to switch frames multiple times during a program, so it is necessary to ensure that the correct frames are selected each time. If one or both of the frames is incorrect, then the position will be in the wrong place and potentially cause a crash. There is no indication in the movement line of these frame settings. This information is found in the Position Detail screen.

Reading Movement Lines

It is important that a programmer understands every detail of a movement line in order to know exactly what actions the robot will take and how to resolve program issues. A programmer must be able to read through a program and identify missing or out-of-place commands. The following is a common movement line:

```
5:J @P[1:Home] 100% FINE
```

Line Numbers. The line number is on the left side of the movement line. Line numbers are set by the software and must stay in order. Lines can be left blank, which causes no problems and may even be helpful for organizing the program. For example, space can be left between different parts of a program so they can be easily found and identified.

Note: When the point is created, the symbol @ will appear between the *J* and *P*. This symbol indicates that the robot is currently at that position.

Movement Types. To the right of the line number is the movement type. A *movement type* is a designation for the shape of the definable motions of a TCP. The possible movement types are J (joint), L (linear), C (circular), and A (circle arc).

Position Information. The *P* in the movement line stands for position. Identifying information for each point is included in the brackets. First is the position ID. A *position ID* is a unique number identifying a particular point in space, along with orientation angles and the associated tool frame number and user frame number. Each unique number represents a specific point in space. There can be up to 32,767 defined positions in a program.

Optionally, the position may also be named so that the programmer can differentiate the defined positions more easily. For example, a position may be `P[1:Home]`. To do this, highlight the position ID number and press ENTER. Then select `Options/Keybd` and then F5 `KEYBOARD` in order to enter a name.

A particular position may be used more than once. That is, many movement lines in different parts of the program may use the same position ID when referring to the same point in space.

More information on a position can be found on a Position Detail screen. To view this information, highlight the position ID and then press F5 `[POSITION]`. **See Figure 4-14.** The position detail data contains the X, Y, and Z positions; W, P, and R orientations; the tool frame; and the user frame information.

Speed. The robot's speed can be specified in a number of ways, including as a percentage (%) of the maximum allowable speed, in the number of seconds (sec) to complete the movement, or in typical speed units, such as inches per minute (inch/min) or milliseconds (msec). It is important to understand that servomotors do not instantaneously start and stop at the specified speeds. There are three phases for each movement: acceleration, constant speed, and deceleration. In certain cases, the robot's time to move from point A to point B may actually be more accurate and faster if the robot's speed is reduced somewhat, which allows for a smoother motion.

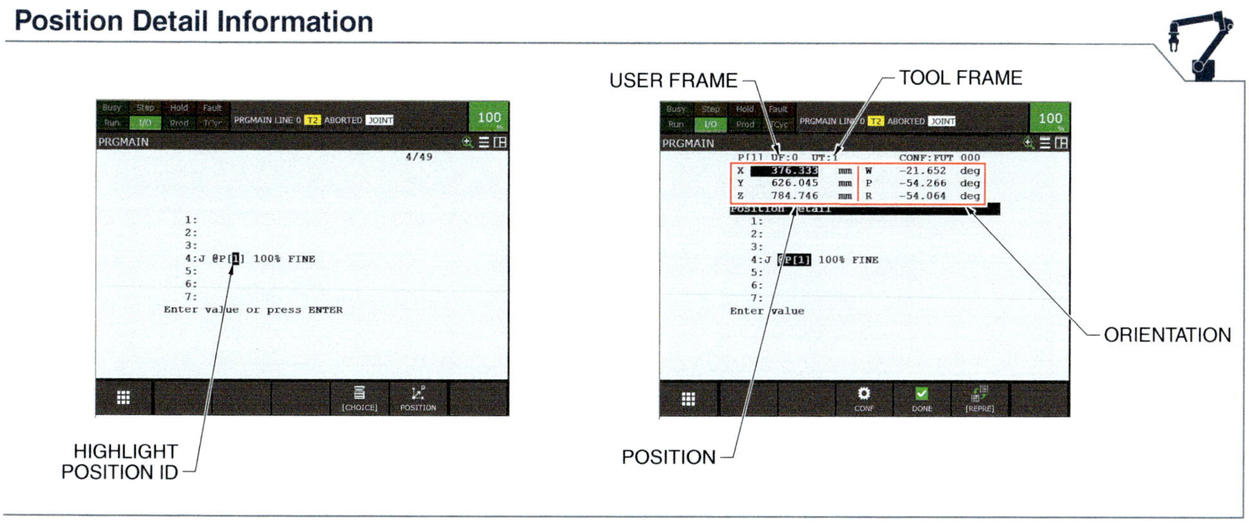

Figure 4-14. Position detail information is accessed by highlighting a position ID and pressing F5 POSITION.

Termination Type. The last item in a movement line is the termination type. A *termination type* is a specification of how accurately the robot passes through a position as it continues to the next position. This determines how smoothly the robot moves through a sequence of movements. The two termination types are fine and continuous (CNT). CNT terminations can also be set between CNT0 (essentially the same as fine) and CNT100 (smooth motion that sacrifices positional accuracy). Termination types will be discussed further in a later chapter.

Re-creating Existing Positions

Having all the information attached to a position number makes it easy to re-create the position in other parts of a program. If the robot must return to a certain position, the programmer can simply create a point (regardless of the current location of the robot, user frame, or tool frame), highlight the position number, and change it to the desired position ID number. This has the effect of changing all position information just created to previously existing position information, including user frame and tool frame numbers.

Touching Up Robot Positions

Sometimes, positions defined in movement lines are not as precise as the programmer desires. There are two methods to adjust a position. Both require first using the arrow buttons to highlight the line number of the position to be adjusted.

The first method is to simply delete the line by using the Delete command in the Edit Command menu. The second way is to use the Touchup command. This is a way to replace the position ID in an existing movement line without re-creating the whole line and its various settings.

If the position ID being replaced is used in more than one movement line, the option is available either to replace all position references with the redefined position or to replace only the line highlighted. If only the one line is changed, then the position is given a new position ID to differentiate it. The other movement lines retain their original position definition. **See the *Using the Touchup Command* Procedure.**

PROCEDURE

Using the Touchup Command

1. Jog the robot to the new position to be recorded.
2. On the teach pendant, use the arrow buttons to highlight the line number of the movement line to be modified. Then press SHIFT and then F5 `Touchup`. (If `Touchup` is not shown, press NEXT first.)
3. The bottom of the screen displays `Set new ID?` Press F4 `YES`. (Selecting F5 `NO` would change all instances of this position ID with the new position definition.)
4. The software places a new position ID number in the brackets to the right of the P.

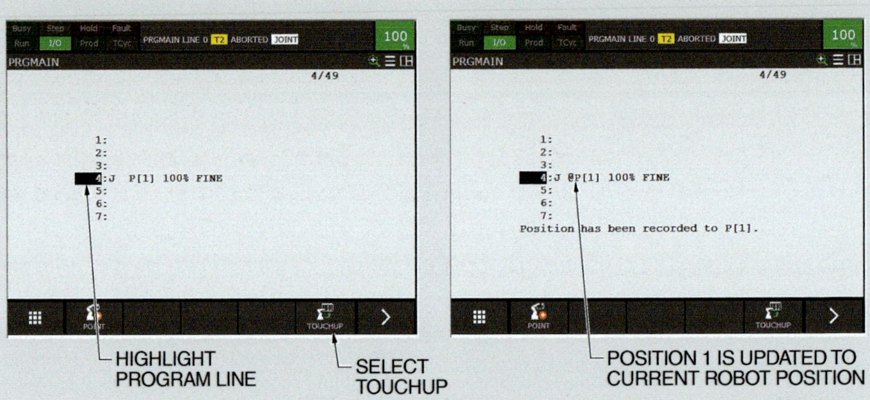

5. If the existing position is in a different user frame than what is currently set in the Jog menu, the bottom of the screen displays `Enter uframe no.`

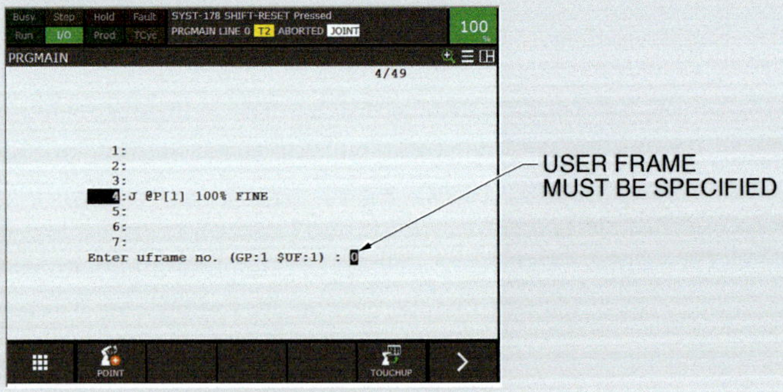

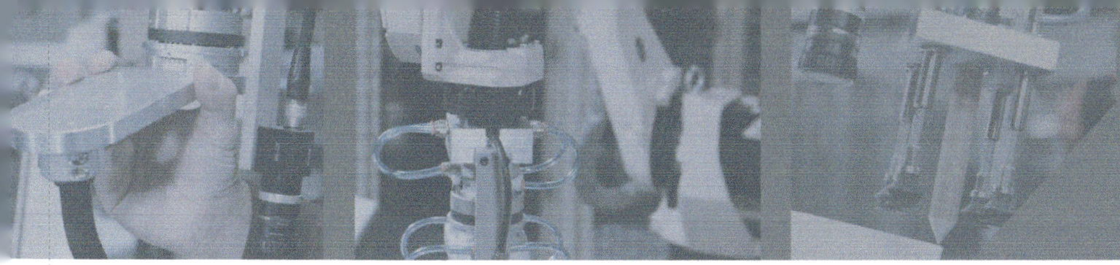

Review 4

Name ______________________________ Date ______________

SECTION 4.1 IDENTIFYING FILE TYPES

_______________ **1.** TP programs consist of ___ lines and command lines.

A. macro

B. movement

C. position ID

D. system

_______________ **2.** ___ provides a precise and specific level of control that TP programming is not capable of, but it is needed in only a few select situations.

_______________ **3.** A(n) ___ file is a computer file containing a list of parameters and their values, which can be adjusted to change the way the teach pendant and robot function.

A. MR

B. TP

C. KAREL

D. SV

4. What are the two types of program files most commonly used?

5. How are TP programs created?

6. What are two advantages to using MR programs?

SECTION 4.2 SETTING UP TP PROGRAMS

_______________ **7.** A program ___ is a program file containing the program's name and empty lines but does not yet contain any code.

_______________ **8.** The ___ parameter defines the type of program file.

A. sub type
B. group mask
C. program detail
D. comment

_______________ **9.** A(n) ___ is a set of axes, such as a robot's axes, and also possibly other electromechanical devices that could be controlled by the robot program.

_______________ **10.** The first position in the group mask signifies the ___.

A. end effector
B. robot
C. first axis
D. pneumatic system

_______________ **11.** When the ignore pause parameter is turned off, a programmer or operator can pause the program by pressing ___.

12. What are the first two steps in creating a TP program?

13. Where would a programmer find the write protect parameter, which prevents other programmers from unintentionally overwriting lines in a program?

SECTION 4.3 CREATING PAYLOAD SCHEDULES

_______________ **14.** A(n) ___ is the combined mass of an end effector and any object that is picked up and handled by the end effector.

______________________ **15.** When the robot knows the mass of a payload, the controller adjusts the robot's ___, giving the servomotors (which run the robot's axes) the ability to stop the robot in the correct position.

A. orientation
B. speed
C. brake pressure
D. path of travel

______________________ **16.** The payload's center of mass and moment of inertia are referenced from the ___, which is the robot's default TCP.

A. center of the end effector
B. line of singularity
C. center of Axis 6
D. World origin

17. Why are multiple payload schedules needed?

SECTION 4.4 USING EDITING COMMANDS

______________________ **18.** When the Insert command is used, the lines inserted ___ the highlighted program line.

A. copy
B. replace
C. appear below
D. appear above

______________________ **19.** The Comment command toggles ___ of comments.

20. What are the first three options for pasting copied program lines?

SECTION 4.5 PROGRAMMING MOVEMENT LINES

____________________ **21.** It is best practice that the ___ position be the first point and movement line in a program.

____________________ **22.** The ___ menu enables the programmer to set both the tool and user frame for the next programmed point.

A. Jog
B. EDCMD
C. Program Detail
D. Frame

____________________ **23.** The frame settings for an existing movement line are found in the ___ screen.

A. Jog
B. EDCMD
C. Program Detail
D. Position Detail

____________________ **24.** A(n) ___ is a unique number identifying a particular point in space, along with orientation angles, tool frame numbers, and user frame numbers.

____________________ **25.** A ___ is a specification for how accurately the robot passes through a position as it continues to the next position.

A. movement type
B. termination type
C. speed
D. command line

____________________ **26.** The ___ command is a way to replace the position ID in an existing movement line without re-creating the whole line and its various settings.

27. What are the benefits of the Home position?

28. Why is it important to verify the tool and user frames for each movement line?

29. What are the four movement type letters and their meanings?

30. How is a point re-created?

4 Lab Exercise

Lab Objective

Create a new program and add code to trace two objects with different user frames.

Equipment Required

- Robot with end effector, pointer, and associated tool frame
- Two square objects, placed in the robot cell, and their associated user frames

Lab Procedure

1. Create a program shell with 50 lines.
    - The program name should be "SquareTrace". (Remember that there can be no spaces or special characters.)
    - The sub type should be None (meaning, a regular TP program).
    - No additional robot groups are required. If the robot is equipped with additional groups, eliminate them by changing the 1s to asterisks (*).
2. In the Jog menu, select User Frame 0, which is the World frame. The tool frame should be one created for the current end effector with its pointer.
3. Create a Home position on line 1. The Home position should be a neutral position for the robot.
4. In the Jog menu, change the user frame number to the user frame for the first square. The tool frame stays the same.
5. Move the cursor down several blank lines and trace the first square. Jog the robot to a corner so that the pointer on the end effector is about ½″ above the square and record the position. Do this for all four corners.
6. Move the cursor to the next blank line and create a fifth position. Change its position number to the number for the first corner of the square.
7. In the Jog menu, change the user frame number to the user frame for the second square. The tool frame stays the same.
8. Move the cursor down several blank lines and trace the second square. Jog the robot to a corner on the second square and create positions on each of the corners as was done on the first square.
9. Move the cursor down several blank lines. Create another position and change the position number to the Home position number.

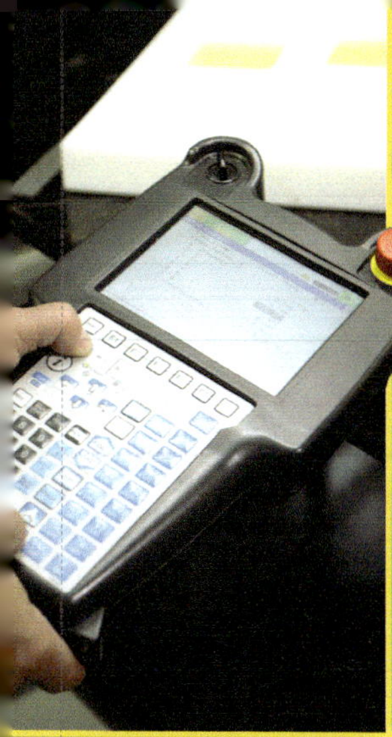

5 Comparing Movement Types

A robot must be able to move in potentially complex paths consisting of arcs, circles, and straight lines. The robot must complete these movements quickly, accurately, and smoothly. When programming these paths, a programmer has four types of movements to choose from. It is the programmer's responsibility to select the proper movement types and speeds in order to achieve the required operating efficiency.

Objectives

SECTION 5.1 SELECTING MOVEMENT SETTINGS

- Describe how movement parameters are applied to programmed points.
- Demonstrate how to change movement type, speed, and termination parameters.
- Describe the considerations when selecting speed.
- Describe the effects of termination types on robot movement.

SECTION 5.2 PROGRAMMING MOVEMENT TYPES

- Explain why joint movement is the fastest of the movement types.
- Explain why linear movements are slower than joint movements.
- Describe considerations for the positions of the points defining a circle.
- Demonstrate how to program a robot to move the end effector in a circle.
- Demonstrate how to program a robot to move the end effector in a circle arc.

Key Terms

- fine termination
- continuous termination (CNT)
- joint movement
- linear movement
- circular movement
- circle arc movement

Learner Resources
atplearningresources.com
Access Code: **762316**

SECTION 5.1 SELECTING MOVEMENT SETTINGS

Robot motion is defined as a series of movements from point to point. Each segment of movement can be customized using certain parameters. Programmers apply these parameter settings to the destination point. **See Figure 5-1.** For example, if the programmer wants the robot to move from point A to point B in a straight line, then point B is programmed with a linear move. Any settings applied to point A define the preceding movement (that ends at point A) and do not affect the movement to point B.

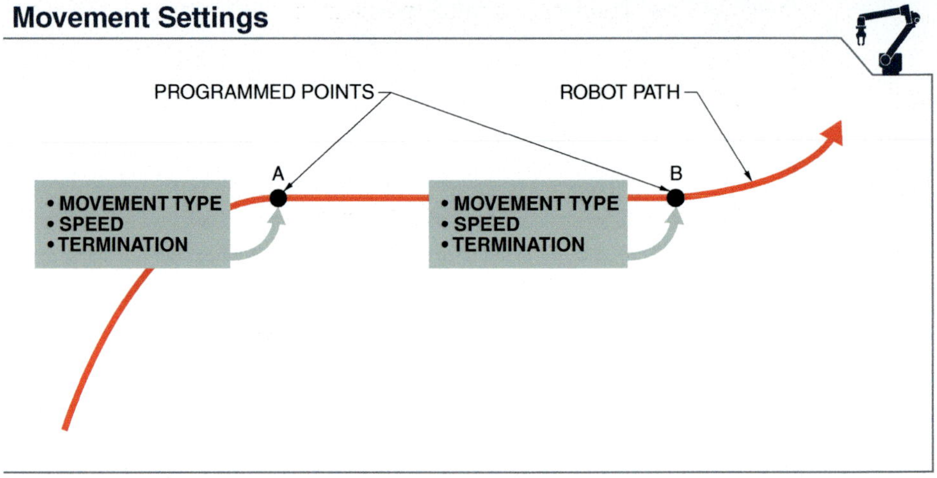

Figure 5-1. The parameter settings of movement type, speed, and termination are applied to the destination point.

Movement Types

The first parameter is movement type, and it determines the type of path the robot takes between points. The options available are joint, linear, circular, and circle arc. **See Figure 5-2.** The choice of movement type affects not only the path but also limits the maximum speed, as some movements are computationally intensive on the controller computer.

Movement types are represented by single letters in the movement line: J for joint, L for linear, C for circular, and A for circle arc. The movement type can be changed by highlighting the movement-type letter and pressing F4 `[Choice]` when it appears. The menu that appears allows different movement types to be selected. **See Figure 5-3.**

Speeds

Another parameter is speed, and it determines how quickly the robot moves between points. The robot uses several different units of speed, including percentage (%) of maximum allowable speed, seconds (sec), millimeters per second (mm/sec), centimeters per minute (cm/min), inches per minute (inch/min), milliseconds (msec), and degrees per second (deg/sec). Not all units are available to use with all movement types though. **See Figure 5-4.** Speed unit options change with the selection of movement type.

The speed unit type can be changed by highlighting the speed and pressing F4 `[Choice]`. A menu appears that allows different speed types to be selected. **See Figure 5-5.** Then the menu provides the option to type the associated number value.

The programmer selects the appropriate speed for the movement type and length of movement the robot will execute. If the selected speed is too high, the servomotors will not be able to meet the acceleration, constant speed, or deceleration requirements, which results in the robot never reaching the selected speed and its movement appearing jerky. This affects the robot's position accuracy and the program's repeatability.

In terms of speed selection, a robot is similar to a car. A car is not designed to run at top speed or to go speeding down a rough logging road. If it does, however, it will result in an increased number of problems and a shorter life expectancy of the car. The same thing will happen to a robot if the programmer pushes it to the maximum speed without concern for the condition of the robot's components.

To correct this situation, the programmer must slow the robot's movement down. This results in smoother robot operation, more accurate positioning, and faster task completion.

Movement Types

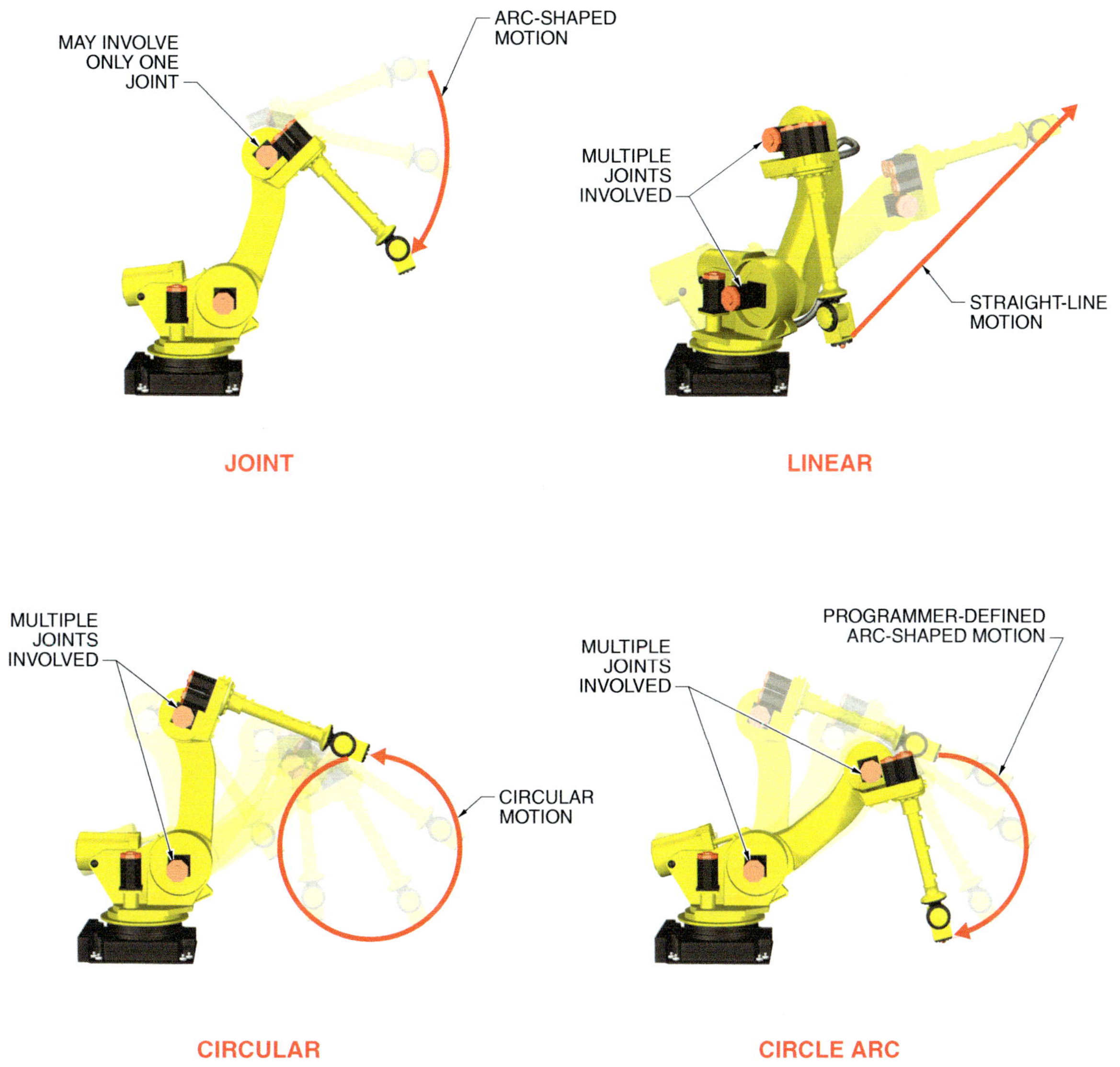

Figure 5-2. The four movement types are joint, linear, circular, and circle arc.

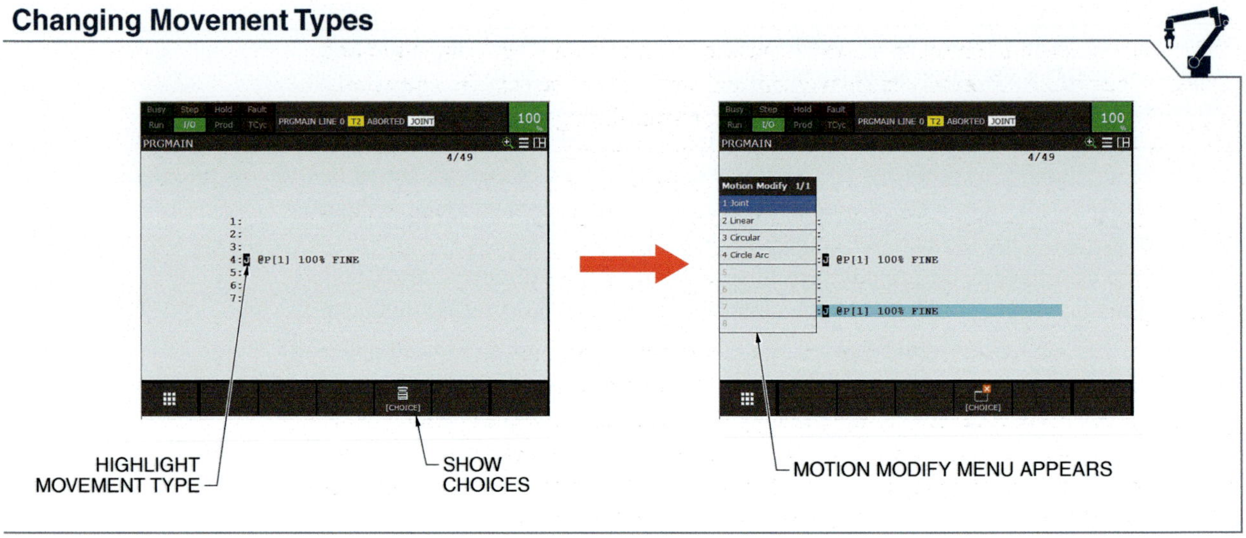

Figure 5-3. Movement types are changed by highlighting the movement-type letter and pressing F4 `[Choice]`.

Available Speed Types						
Speed Type	**FANUC Abbreviation**	**Range**	**Joint**	**Linear**	**Circular**	**Circle Arc**
Percentage of maximum speed	%	1 to 100	✓			
Seconds	sec	0.1 to 3200	✓	✓	✓	✓
Millimeters per second	mm/sec	1 to 2000		✓	✓	✓
Centimeters per minute	cm/min	1 to 12,000		✓	✓	✓
Inches per minute	inch/min	0.1 to 4724.4		✓	✓	✓
Milliseconds	msec	1 to 32,000	✓	✓	✓	✓
Degrees per second	deg/sec	1 to 630		✓	✓	✓

Figure 5-4. There are several speed types, but not all are available for every movement type.

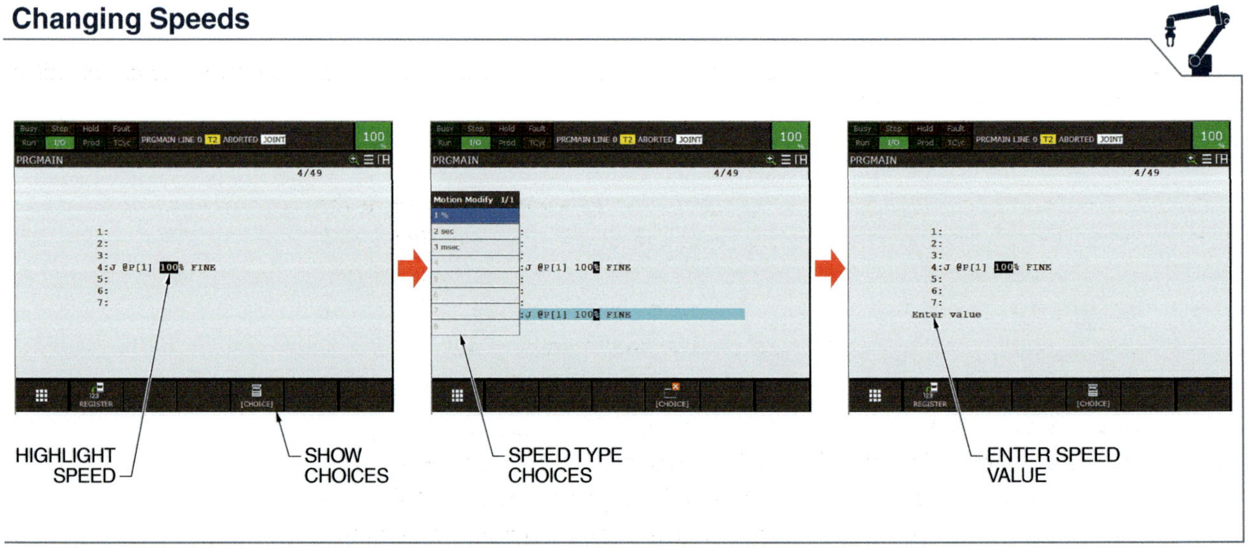

Figure 5-5. Speed type is changed by highlighting the speed and pressing F4 `[Choice]`. The numerical value is changed by highlighting the current value and typing a new one.

Terminations

The last parameter in the movement line is position termination. The termination of a point describes how a robot ends one movement and continues to the next. In other words, the termination type determines how closely a robot moves through a programmed point.

A programmer does not always require a robot to move precisely through a point. For example, a pencil is on a table. A person wants to pick the pencil up, but the person's hands are in their pockets. The person is not concerned with the exact path their hand takes when reaching for the pencil as long as they do not hit the table before reaching it. For the most part, when people move their hands, it is a fluid, rounded movement, not jerky and squarish.

To create similar smooth, fluid movements in a robot, a programmer uses different types of terminations throughout a robot program. There are two types of terminations: fine and continuous (CNT). The termination type can be changed by highlighting it and pressing F4 `[Choice]` when it appears. **See Figure 5-6.** A menu appears that allows either fine or CNT termination types to be selected. If CNT is selected, the number after CNT is highlighted, and a new number can be entered with the keypad.

Changing Termination Types

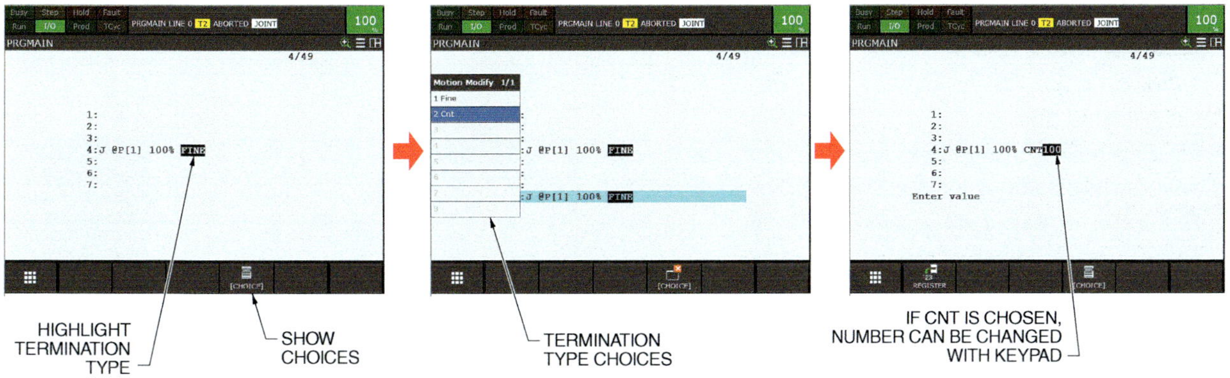

Figure 5-6. A termination type is changed by highlighting the type and pressing F4 `[Choice]`. If CNT is chosen, an option to add the associated numerical value appears.

Fine Terminations. A *fine termination* is an instruction to a robot to move to the programmed position, stop, and then accelerate to the next position. **See Figure 5-7.** A fine termination is often used for the position just before gripping an object. In the pencil example, the person's hand moves down toward the pencil but stops just above it. They then close their fingers around the pencil. The programmer would use the fine termination on the downward movement just before touching the table.

Continuous Terminations. A *continuous termination (CNT)* is an instruction to a robot to allow a certain amount of deviation from a defined point in order to move smoothly through it to the next point. The robot moves toward a CNT-programmed point, decelerates, and then accelerates to the next point without actually reaching the programmed point. In the pencil example, the movement of the hand from the pocket toward the pencil would be continuous type.

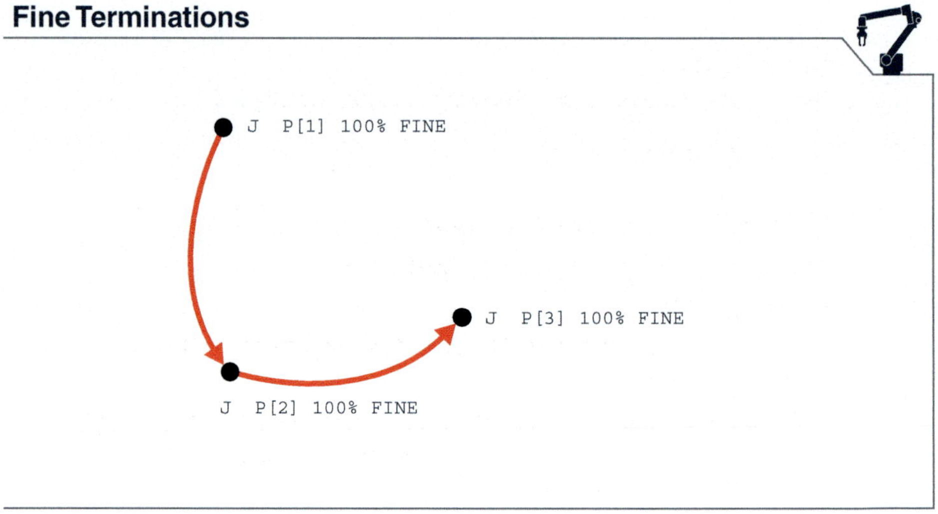

Figure 5-7. Fine terminations pass exactly through programmed points. This is precise, but it requires the robot to slow down and speed up repeatedly.

Robot Remastering

Each servomotor must be zeroed to a reference point. This is how the controller identifies the position and orientation of the robot, sends axis limit faults (alarms), and ensures precision and repeatability. This zero point is visible as a reference mark aligned to the center of the witness mark. The precise zeroing to the reference mark is called mastering. A hard crash, loss of all power (including battery backup), or motor cable disconnection can cause a loss of axis mastering. One important way to prevent the loss of axis mastering is to change the batteries every six months.

If a robot loses its mastering, the fault will be SRVO-062 BZAL or SRVO-038 BZAL. These faults indicate that a servomotor is not located where the controller is expecting it to be. After resetting the fault, the robot should be placed in its zero position (according to the witness marks), and the "quick mastering" procedure is used to recalibrate the zero reference.

Remastering the Robot

Reset Fault (Alarm)

1. Press MENU.
2. Use the arrow buttons to highlight `Alarm` and then `Alarm Log`. Then press ENTER.
3. Press F4 `RESET`.

Position the Robot

4. Jog each axis of the robot until the witness marks are aligned.

Select Remastering Type

5. Press MENU.
6. Use the arrow buttons to highlight -- `NEXT` -- and press ENTER.
7. Use the arrow buttons to highlight `SYSTEM` and then `Variables`. Then press ENTER.
8. Press F1 `[TYPE]`.
9. Use the arrow buttons to highlight `Master/Cal` and press ENTER. If `Master/Cal` is not shown, then enable this function:
   a. Exit the Type menu by touching outside the menu or pressing PREV.
   b. Use the arrow buttons to scroll down to the variable `$MASTER_ENB` and press ENTER.
   c. Type 1 on the keypad and press ENTER.

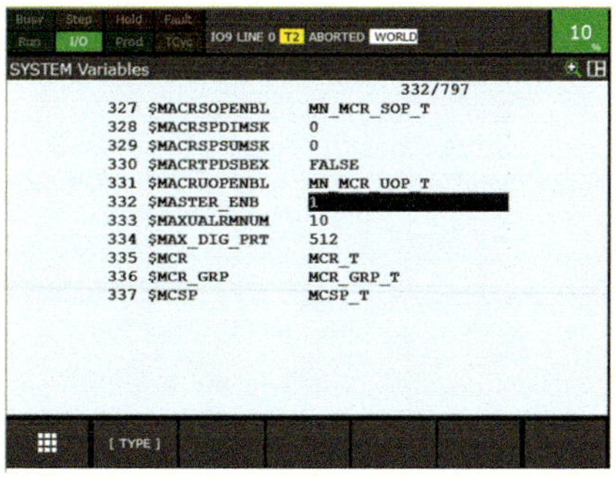

   d. Return to the Type menu, select `Master/Cal`, and press ENTER.
10. Select `QUICK MASTER` and press ENTER.

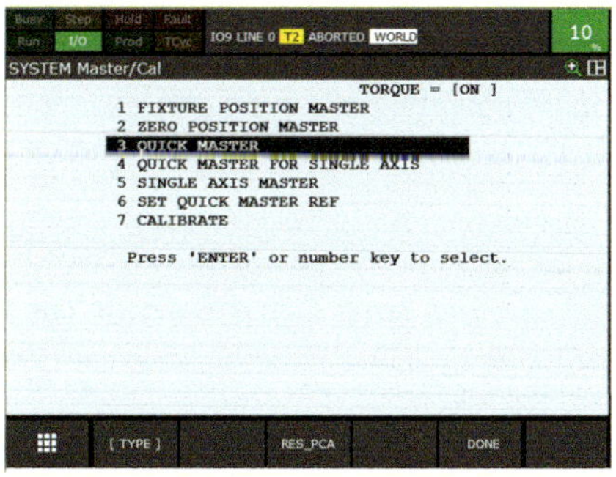

11. Press F4 `Yes`.

Cycle Robot Power

12. Press FCTN.
13. Use the arrow buttons to highlight -- `NEXT` -- and press ENTER.
14. Use the arrow buttons to highlight `CYCLE POWER` and press ENTER. The screen displays:
    `This will cycle power. Are you sure?`
15. Select `YES`.

The amount of deviation is specified by a number from 1 to 100. A setting of CNT1 is nearly the same as a fine termination. That is, the robot comes very close to the point before accelerating toward the next point. **See Figure 5-8.** As the number is set higher, the robot is allowed to be farther from the programmed point. The numbers 1 to 100 are arbitrary numbers. There is no specific distance equated to them.

A movement using continuous termination is faster than a movement using fine termination. In addition, a termination of CNT100 is faster than a CNT with a lower number. When a programmer uses a combination of fine and continuous movements, and carefully chooses CNT numbers, the robot will be able to move in a fluid, nonjerky motion. **See Figure 5-9.**

When deciding what number to use with a CNT, the programmer must rely on experience. There is no rule of thumb regarding how close a robot can or should come to the programmed position or how much of an arc needs to be created when traveling to the position.

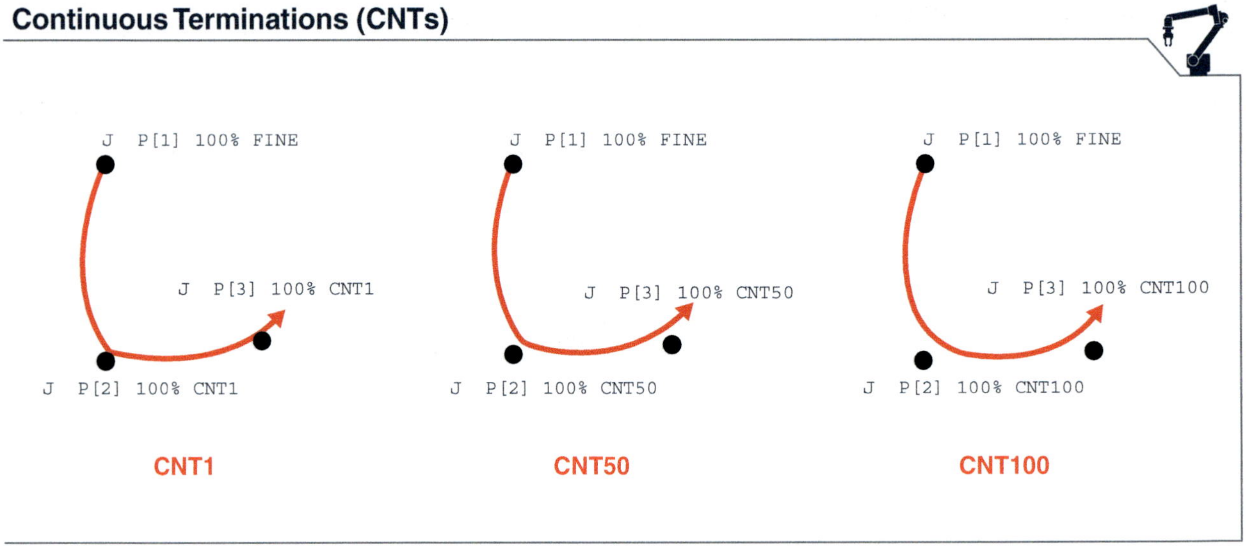

Figure 5-8. CNTs are useful when it is more important to have a smooth motion than to pass precisely through a point. CNT values determine how closely the TCP must pass through programmed points.

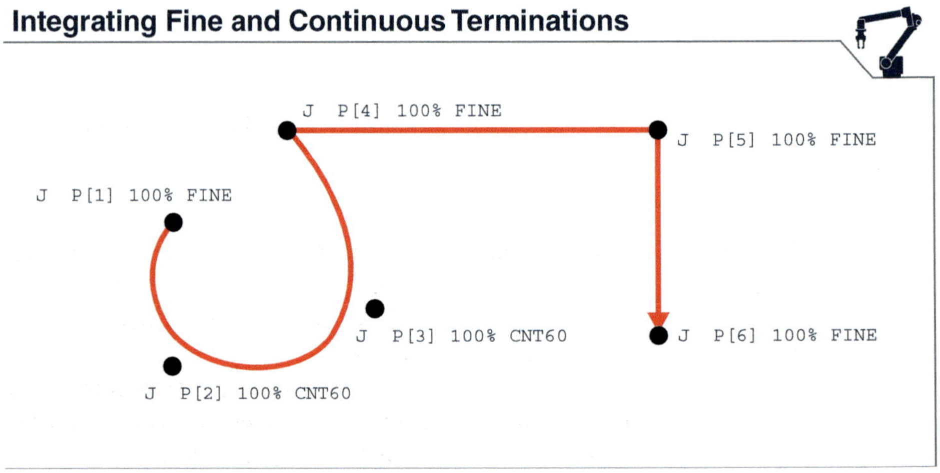

Figure 5-9. Fine terminations and CNTs are used in a program together to balance fast and smooth movements with accurate positioning.

SECTION 5.2 PROGRAMMING MOVEMENT TYPES

The four different movement types are joint, linear, circular, and circle arc. These movement types determine how the robot's TCP moves between defined points.

Joint Movements

Joint movement is a type of movement in which the TCP is moved with the most efficient joint rotations. This is generally the fastest movement type because the controller needs very little processor time to determine the necessary joint rotations. Each joint is moved as needed for the TCP to reach the destination point, without consideration to the path taken or the orientation of the end effector. Each joint rotation moves the TCP in an arc. Even if the movement requires multiple joints to rotate, the path from point A to point B in joint movement will always be an arc. **See Figure 5-10.**

Joint movement only has three selections for speed: percentage (%), seconds (sec), and milliseconds (msec). It is the only movement type that has percentage as a speed selection.

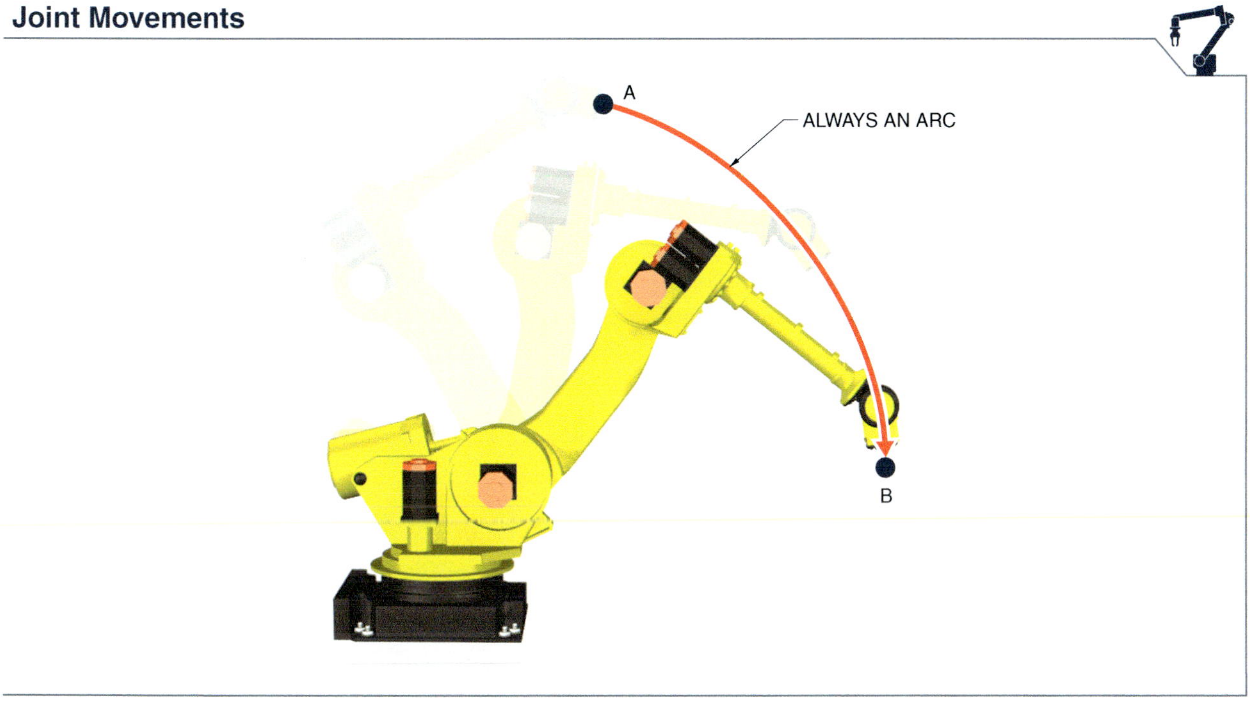

Figure 5-10. Joint movement is efficient and fast. It always moves the TCP in an arc.

Linear Movements

Linear movement is a type of movement in which the TCP moves in a straight line. **See Figure 5-11.** Each millimeter of linear movement requires the controller computer to recalculate the position of each axis. Since these calculations are complicated, the controller responds slower. Linear movements are slower than joint movements.

Speed selection for a linear movement also differs from that of a joint movement. The percentage (%) speed option is not available for a linear movement. In linear movement, the programmer's choices for speed selection are the following: millimeters per second (mm/sec), centimeters per minute (cm/min), inches per minute (inch/min), degrees per second (deg/sec), seconds (sec), and milliseconds (msec).

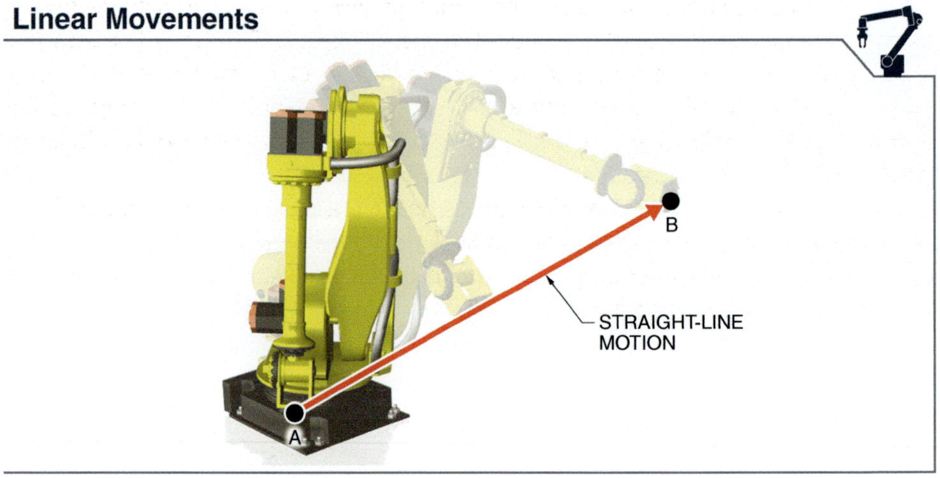

Figure 5-11. Linear movements are in a straight line.

Circular Movements

Circular movement is a type of movement in which the TCP moves in a full circle. First, the programmer must envision the desired circle broken up into four equal parts defined by four points. **See Figure 5-12.** The first point has settings for the movement coming into that point, and those settings may vary. The second and fourth points are initially defined as joint by default but are changed to circular movement type. This has two effects. First, an additional line appears below each defined point. These placeholders are then respectively assigned the positions of the third and first points. Assigning the first point at the end of the circle closes the circular motion and returns the robot to the starting point. Second, the speed is changed to millimeters per second (mm/sec) since a percentage speed type is not applicable to circular movements. However, this can then be changed to seconds (sec), centimeters per min (cm/min), inches per minute (inch/min), milliseconds (msec), or degrees per second (deg/sec). **See the *Creating a Circle* Procedure.**

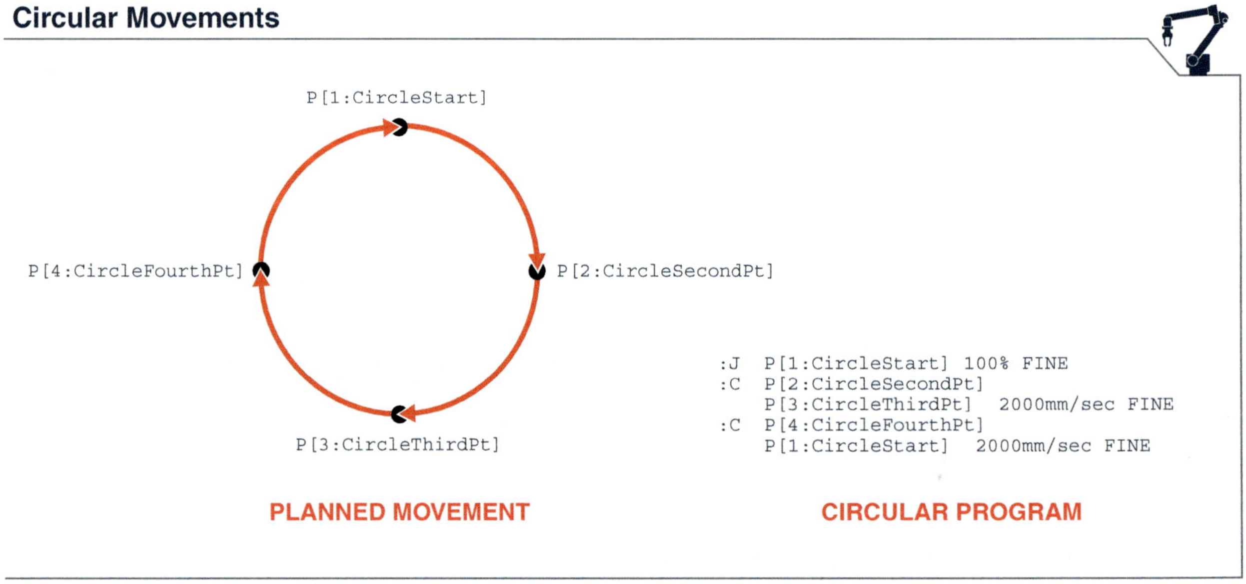

Figure 5-12. Circular movements are based on four points defined on a planned circle.

It is, of course, difficult to define the points precisely to make a mathematically perfect, two-dimensional circle in three-dimensional space. The controller uses the programmer-defined points and fits the robot motion to them, even if it results in a "wavy circle" or an ellipse, but only to a degree. If the defined points stray too far from a roughly circular shape with fairly equidistant points, the controller will enter a fault condition.

The speed selections and limitations for a circular movement are the same as those for a linear movement. Like linear movement, the controller must intensively calculate the robot's joint rotations to keep the end effector traveling on the precisely defined circular path. This limits the speed.

Circle Arc Movements

The fourth type of movement is a circle arc. *Circle arc movement* is a movement in which the TCP moves in arc-shaped paths. A programmer might use a circle arc movement type to ensure that the robot avoids colliding with an object in the work cell or to trace a half-round item. For example, robots that apply adhesives or weld may need to move in an arc in order to precisely trace an object.

While both joint and circle arc movement types move the robot in arcs, there is a big difference between them. A joint movement type follows whichever arc path is the most efficient (least number of axis moves) to move between two points. A circle arc movement type, however, defines a precise arc-shaped path between points that must be followed and which may involve multiple axes.

Three points are required to make an arc. Each point is first defined in the default joint movement type and then changed to the circle arc movement type. **See Figure 5-13.** The limitation for a circle arc is about half a circle. An arc longer than that must instead be created with a circular movement. The speed selection for a circle arc movement is the same as for the circular and linear movement types: millimeters per second (mm/sec), centimeters per minute (cm/min), inches per minute (inch/min), degrees per second (deg/sec), seconds (sec), and milliseconds (msec). **See the *Creating a Circle Arc* Procedure.**

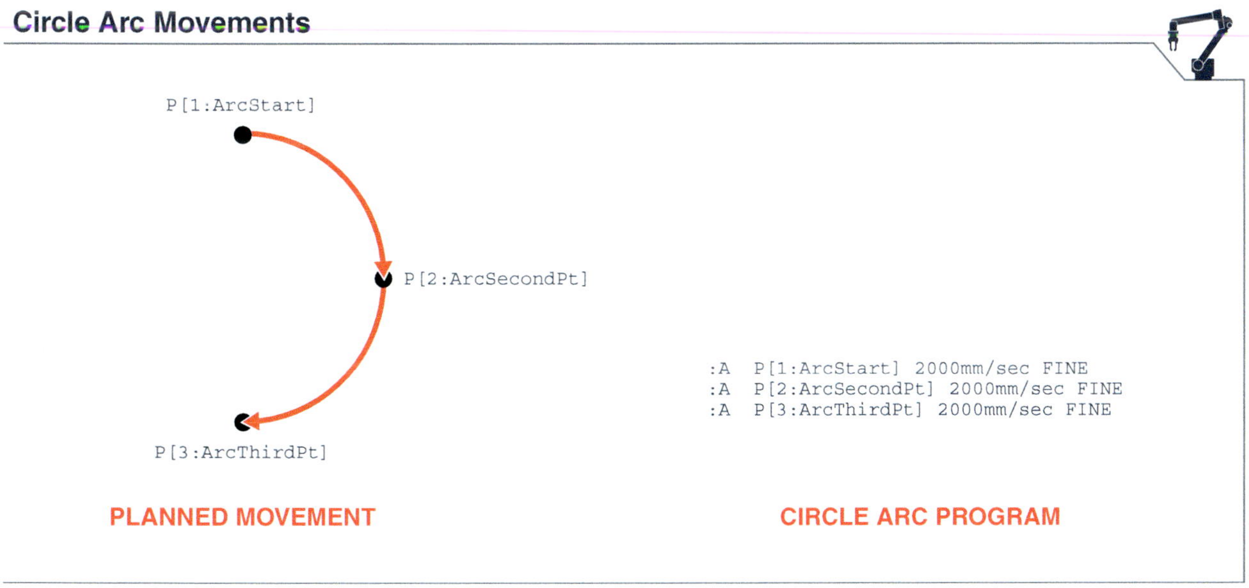

Figure 5-13. Circle arc movements are based on three points.

PROCEDURE

Creating a Circle

Set Up Point 1

(1) Jog the robot to the position where the circle will start.

(2) Press and hold SHIFT and then press F1 `POINT` to establish the first point of the circle. Give the point a descriptive name:

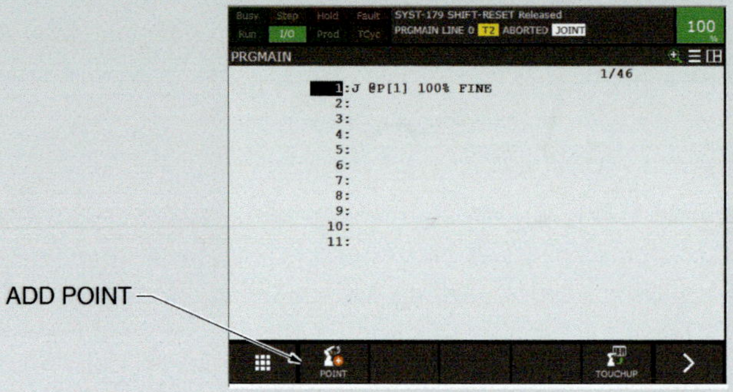

a. Use the arrow buttons to highlight the position ID number and press ENTER to show the `Alpha Input` menu.

b. Use the arrow buttons to highlight `Options/Keybd` and press ENTER.

c. Press F5 `KEYBOARD`.

d. Type in the desired point name, such as "CircleStart," and press EXIT.

Set Up Point 2

(3) Jog the robot to the second position of the circle.

(4) Press and hold SHIFT and then press F1 `POINT`. Give the point a descriptive name, such as "CircleSecondPt." Currently, the point is a joint move.

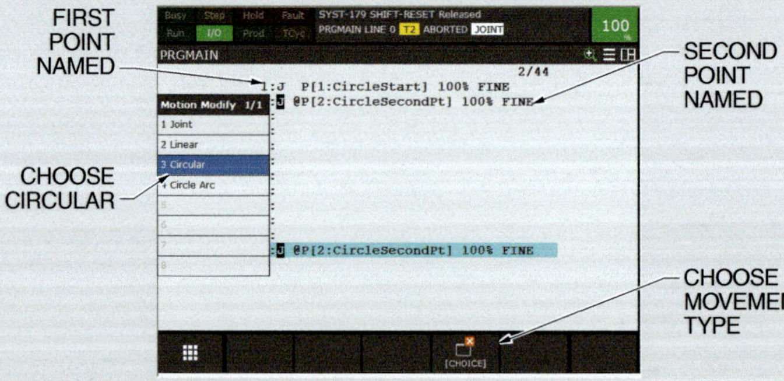

(5) Use the arrow buttons to highlight the `J` of the second point. Then press F4 `[CHOICE]` to show the `Motion Modify` menu.

(6) Use the arrow buttons to highlight `Circular` and press ENTER. This changes the `J` to a `C`, and a new, indented line appears.

(7) Use the arrow buttons to highlight the indented line.

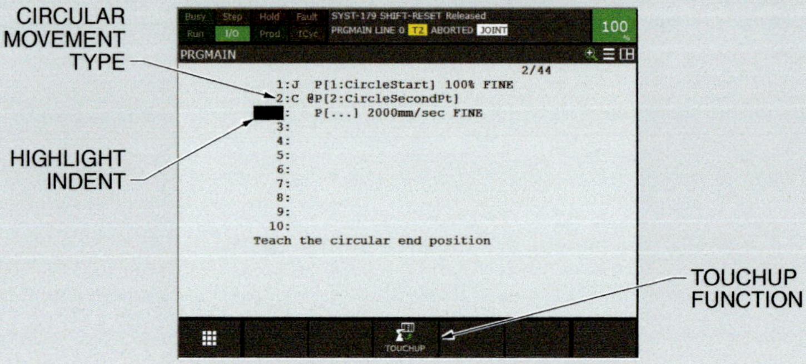

Set Up Point 3

(8) Jog the robot to the third position of the circle.

(9) Press and hold SHIFT and then press F3 `TOUCHUP`. The indented line is automatically assigned a position ID. (A line number is still not assigned to this point.) Give the point a descriptive name, such as "CircleThirdPt." Note that if the line is too long for the programming display, it will automatically wrap to a second line.

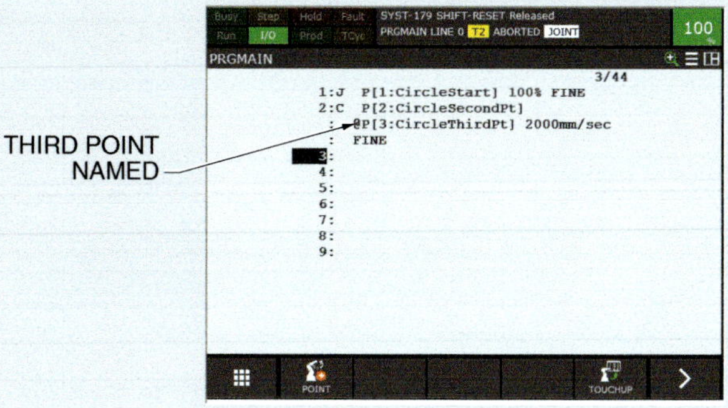

Set Up Point 4

(10) Jog the robot to the circle's fourth position.

(11) Press and hold SHIFT and then press F1 `POINT`. Give the point a descriptive name, such as "CircleFourthPt."

(12) Use the arrow buttons to highlight the `J` of the fourth point. Then press F4 `[CHOICE]` to show the `Motion Modify` menu.

(13) Use the arrow buttons to highlight `Circular` and press ENTER. This changes the `J` to a `C`, and a new indented line appears.

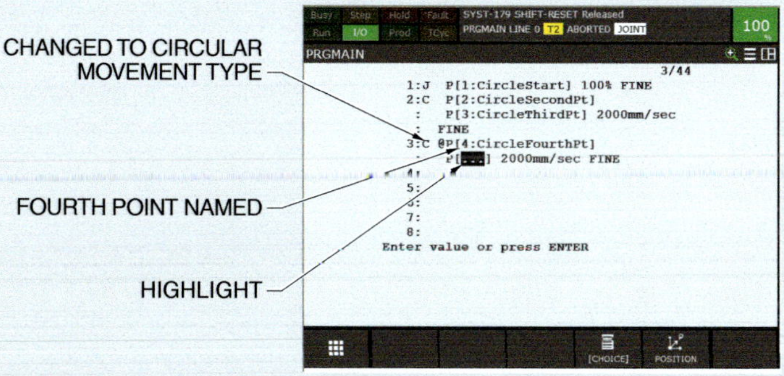

Return to Point 1

(14) Use the arrow buttons to highlight the dots between the brackets.

(15) Enter the position ID of the circle's first point and press ENTER.

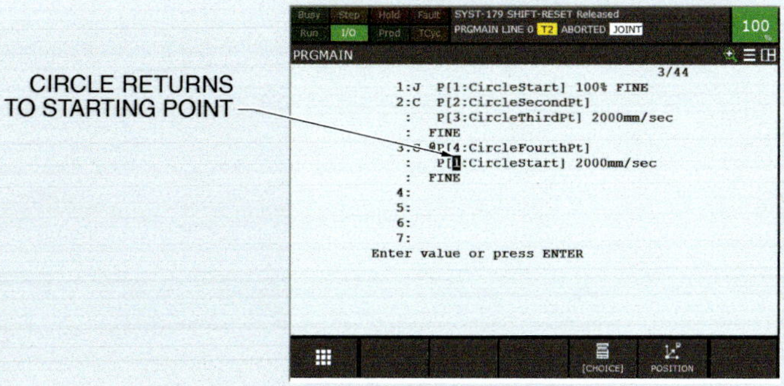

PROCEDURE

Creating a Circle Arc

Set Up Point 1

1. Jog the robot to the position where the circle arc will start.
2. Press and hold SHIFT and then press F1 `POINT` to establish the first point of the arc. Give the point a descriptive name, such as "ArcStart."
3. Use the arrow buttons to highlight the `J` of the first point. Then press F4 `[CHOICE]` to show the `Motion Modify` menu.
4. Use the arrow buttons to highlight `Circle Arc` and press ENTER. This changes the `J` to an `A`. Note that unlike a circular movement, a circle arc movement does not create an indented line.

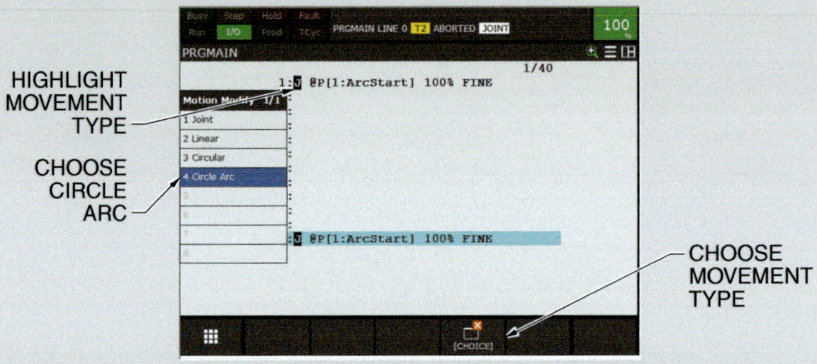

Set Up Point 2

5. Jog the robot to the second position for the arc.
6. Press and hold SHIFT and then press F1 `POINT` to establish the second point of the arc. Give the point a descriptive name, such as "ArcSecondPt."
7. Use the arrow buttons to highlight the `J` of the second point. Then press F4 `[CHOICE]` to show the `Motion Modify` menu.
8. Use the arrow buttons to highlight `Circle Arc` and press ENTER. This changes the `J` to an `A`.

Set Up Point 3

9. Jog the robot to the third and final position for the arc.
10. Press and hold SHIFT and then press F1 `POINT` to establish the third and final point of the arc. Give the point a descriptive name, such as "ArcThirdPt."
11. Use the arrow buttons to highlight the `J` of the third point. Then press F4 `[CHOICE]` to show the `Motion Modify` menu.
12. Use the arrow buttons to highlight `Circle Arc` and press ENTER. This changes the `J` to an `A`.

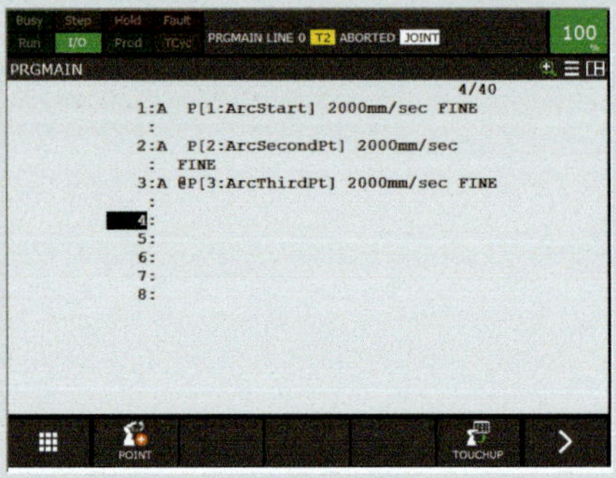

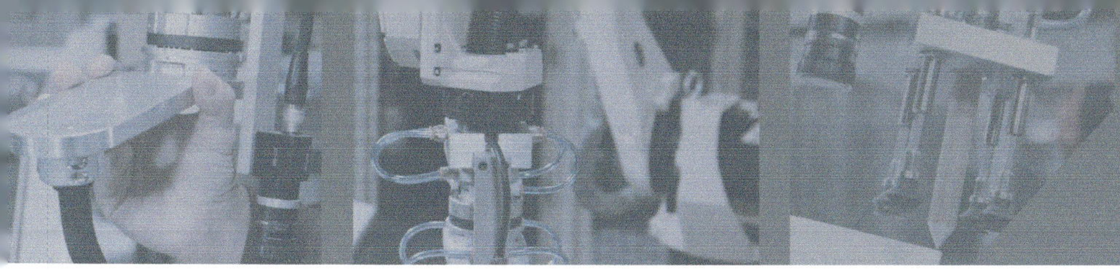

Review 5

Name ______________________________ Date ______________

SECTION 5.1 SELECTING MOVEMENT SETTINGS

______________ **1.** Programmers apply movement parameter settings to the ___ point.

______________ **2.** The ___ type determines the type of path the robot takes between points.

A. termination
B. speed
C. movement
D. continuous

______________ **3.** The ___ type determines how closely the robot moves through a programmed point.

A. termination
B. speed
C. movement
D. fine

______________ **4.** A(n) ___ termination is often used for the position just before gripping an object.

______________ **5.** The amount of deviation for a continuous termination is specified ___.

A. by the speed
B. as a percentage
C. in millimeters
D. with a number from 1 to 100

6. How does speed affect a robot's position accuracy and the program's repeatability?

SECTION 5.2 PROGRAMMING MOVEMENT TYPES

______________ **7.** ___ movement is generally the fastest movement type.

____________________ **8.** The path from point A to point B in a joint movement will always be a(n) ___.

A. straight line
B. arc
C. circle
D. zigzag

____________________ **9.** ___ movement is the only movement type that has percentage as a speed selection.

____________________ **10.** The ___ speed option is not available for a linear movement.

____________________ **11.** A circle program is defined by ___ points.

____________________ **12.** The speed selections and limitations for a circular movement are the same as those for a ___ movement.

A. fine
B. continuous
C. joint
D. linear

____________________ **13.** ___ points are required to make an arc.

14. Why is linear movement slower than joint movement?

15. What is the effect of changing a joint movement type to a circular movement type?

16. What if four programmed points do not define at least a rough circle?

17. What is the difference between the arc motion of a joint movement and a circle arc movement?

18. What is the limitation of a circle arc?

5 Lab Exercise

Lab Objective

Create a new macro program and add code for moving the end effector in a circle.

Equipment Required

- Robot with end effector, pointer, and associated tool frame
- Two square objects, placed in the robot cell, and their associated user frames

Lab Procedure

1. Create a program shell with 25 lines.
    - The program name should be "Circle".
    - The sub type should be Macro.
    - No additional robot groups are required. If the robot is equipped with additional groups, eliminate the groups by changing the 1s to asterisks (*).
2. In the Jog menu, select user frame 0, which is the World frame. The tool frame should be the one created for the current end effector with its pointer.
3. Create a Home position on line 1. The Home position should be a neutral position for the robot.
4. Create a circle in the air over the first square object using the circular movement type. Use the square as a guide for the circle shape, starting at a point in the middle of one side.

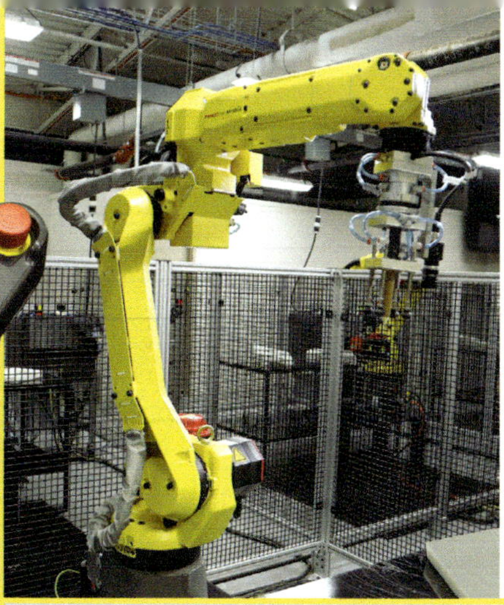

6 Instruction Lines

The Instruction menu includes several robot commands for setting up and directing the execution of program code. These instruction lines do not result in any positioning or movement of a robot, but some are required for robot operation, and others provide important additional features. In many cases, the skilled use of various instruction lines helps organize programs efficiently and optimize cycle times.

Objectives

SECTION 6.1 SELECTING FRAMES

- Explain why the tool and user frame selection lines are needed.
- Describe where frame selection lines are needed.
- Demonstrate how to add tool and frame selection lines.

SECTION 6.2 USING THE PROGRAM ADJUST PROCEDURE

- Describe when to use the program adjust procedure.
- Demonstrate how to create a program adjust schedule and use it to make movement line changes.

SECTION 6.3 DIRECTING PROGRAM EXECUTION

- Describe the purpose of the Call instruction.
- Demonstrate how to call a subprogram.
- Explain the cause of a Stack Overflow fault and how to prevent it.
- Explain how the jump label instruction works.
- Demonstrate how to add jump labels to a program.

SECTION 6.4 FINE-TUNING PROGRAM TIMING

- Explain how timers are useful.
- Demonstrate how to set up a timer in a program.
- Explain why a timer must be reset at the beginning of a program.
- Explain the action of a robot when a program reaches a Wait instruction.
- Demonstrate how to add a wait instruction into a program.

SECTION 6.5 ADDING PROGRAM NOTES

- Explain why remarks are used.
- Demonstrate how to add a remark in a program.
- Explain why messages are used.
- Demonstrate how to add a message in a program.

Key Terms

- program adjust
- program adjust schedule
- subprogram
- jump label
- cycle time
- timer
- remark
- message
- user screen

Learner Resources
atplearningresources.com
Access Code: **762316**

SECTION 6.1 SELECTING FRAMES

Prior to creating a movement line, the programmer decides which tool and user frames to use and selects them in the Jog menu (press and hold SHIFT, and then press COORD). When the next position is created, it then references the frames selected.

Each time the programmer creates a position with a new frame (not the frame used in the previous movement), an instruction line to select the frame is required. **See Figure 6-1.** Failing to place this instruction line results in an Invalid Uframe (user frame) or Invalid Tframe (tool frame) fault. This fault only occurs when the program contains either more than one tool frame or two or more user frames in addition to the World frame.

Frame Selection Lines

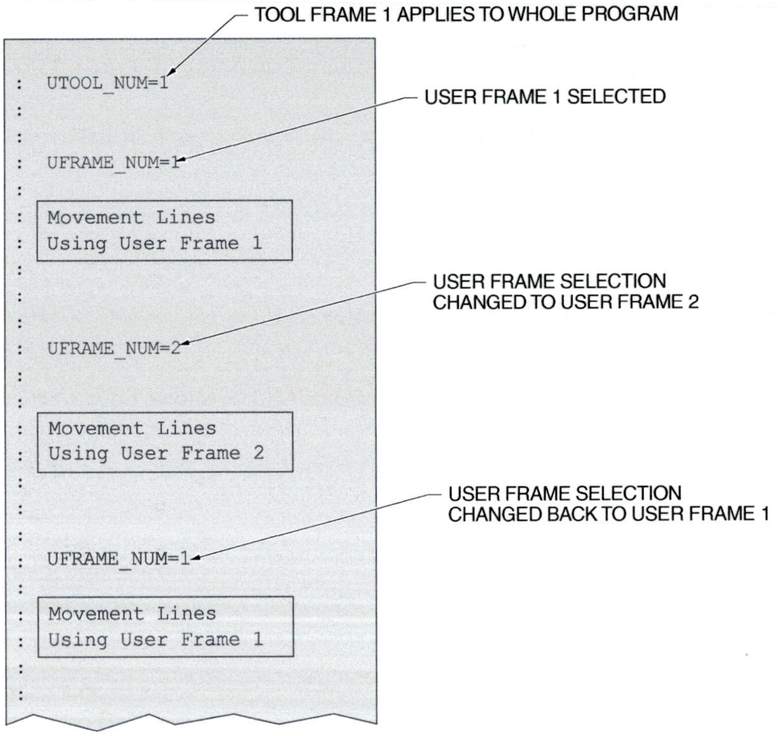

Figure 6-1. Frame selection lines must be included each time a movement line uses a different tool or user frame than the previous movement line.

Even though each recorded point contains information on the tool and user frame associated with it, changing frames during the execution of a program requires these specific declarations. This is a safety feature to ensure the programmer is clear about which frames to use.

If a program uses multiple tool or user frames, it is best to first run the program in manual mode to expose possible errors. This mode moves the robot at a slow speed and allows the operator to immediately stop the robot by simply releasing the SHIFT button. **See the *Running a Program in Manual Mode* Procedure.**

PROCEDURE

Running a Program in Manual Mode

(1) Turn on the teach pendant.
(2) Turn the controller key to T1.
(3) If the fault light is red, reset it by holding the deadman switch in the proper position and pressing [RESET].
(4) Move the cursor to the first line of the program to be run.
(5) Use the speed button to reduce the robot speed to 20%.
(6) While holding the deadman switch and [SHIFT], press and release [FWD].

Resetting Gate Faults

A robot enclosure should have a gate to allow operators to enter. These gates must be equipped with a safety latch, which is wired back to the robot controller. This safety latch places the robot in an E-stop condition if the gate is opened.

A safety gate switch has two redundant channels. If the channel signals both indicate that the gate is closed, then operation continues normally. If both indicate that the gate is open or if the two channel signals do not match, then an E-stop is triggered. The situation wherein the two channels do not match can happen simply by the gate being slammed, from the gate being leaned on, or by the gate switch falling out of alignment. The error that will appear is SRVO-230 Chain 1 Abnormal.

The first step to resetting this fault is to inspect the gate safety latch and wiring to ensure the latch is properly aligned and the wiring has not broken or been cut. If the latch is properly aligned and wiring is in good condition, follow the procedure to reset the fault.

Reset Fault (Alarm)

1. Press [MENU].
2. Select `Alarm` and then `Alarm Log`. Then press [ENTER].
3. Press [F4] `RES_1CH`. The screen displays: `Reset Single Channel Fault?`
4. Press [F4] `YES`.

Cycle Robot Power

5. Press [FCTN].
6. Use the arrow buttons to select -- `NEXT` -- and press [ENTER].
7. Use the arrow buttons to select `CYCLE POWER` and press [ENTER]. The screen displays: `This will cycle power. Are you sure?`
8. Select `YES`.
9. Allow a few minutes for the robot to shut down and reboot. Press no buttons on the teach pendant during this time.

The instruction lines to select the desired tool frame and user frame must be put in the proper positions within the program. If only one tool or user frame is used for the entire program, the instruction lines are put at the beginning of the program, before any movement lines. If a program uses more than one tool or user frame, the corresponding frame instruction line is put before each movement line that changes frames. **See the *Inserting a Tool Frame Selection Line* and *Inserting a User Frame Selection Line* Procedures.**

PROCEDURE

Inserting a Tool Frame Selection Line

① Highlight the desired blank line.

② Press F1 `[INST]`.

③ Highlight --`next page`-- and press ENTER.

④ Highlight `Offset/Frames` and press ENTER.

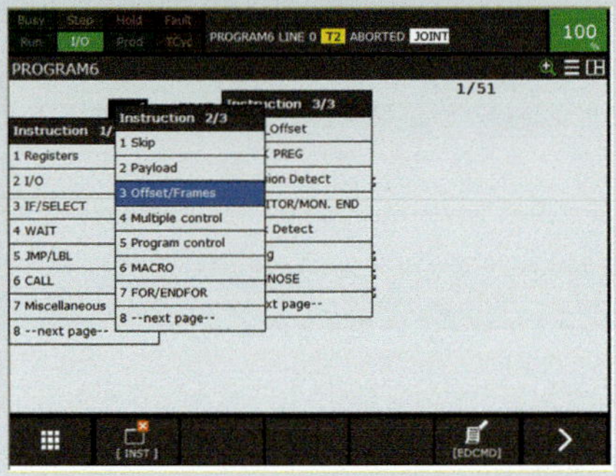

⑤ In the menu that appears, highlight `UTOOL_NUM=...` and press ENTER. The line `UTOOL_NUM=...` is added to the blank line.

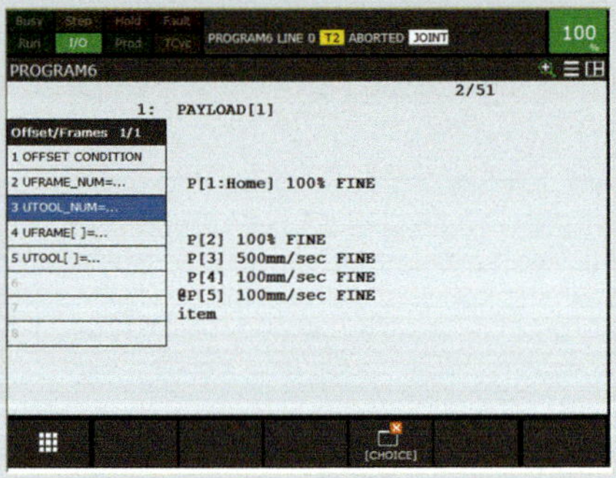

⑥ In the menu that appears, highlight `Constant` and press ENTER. The dots in the `UTOOL_NUM=...` are highlighted.

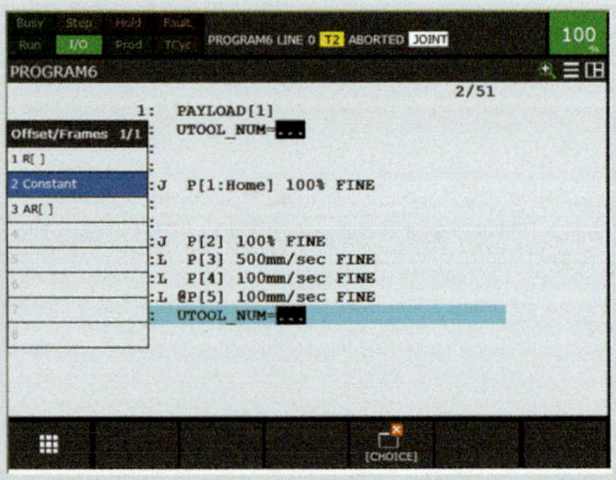

⑦ Using the teach pendant keypad, type the number of the desired tool frame and press ENTER. In this case, the desired tool frame is 1.

8. If needed, repeat the procedure to add a tool frame selection line before each movement line that uses a different tool frame.

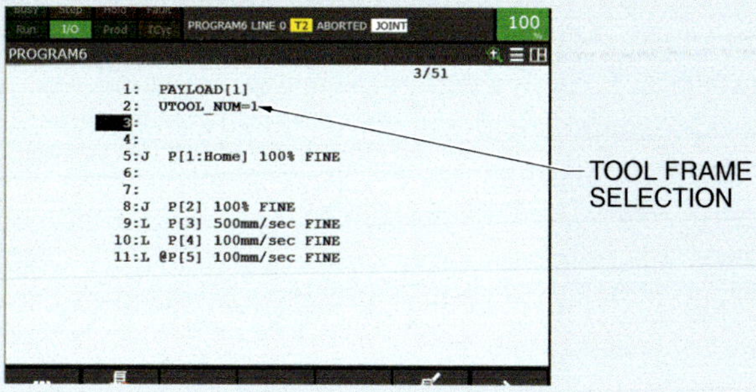

PROCEDURE

Inserting a User Frame Selection Line

1. Highlight the desired blank line.
2. Press F1 [INST].
3. Highlight --next page-- and press ENTER.
4. Highlight Offset/Frames and press ENTER.
5. In the menu that appears, highlight UFRAME_NUM=... and press ENTER. The line UFRAME_NUM=... is added to the blank line.

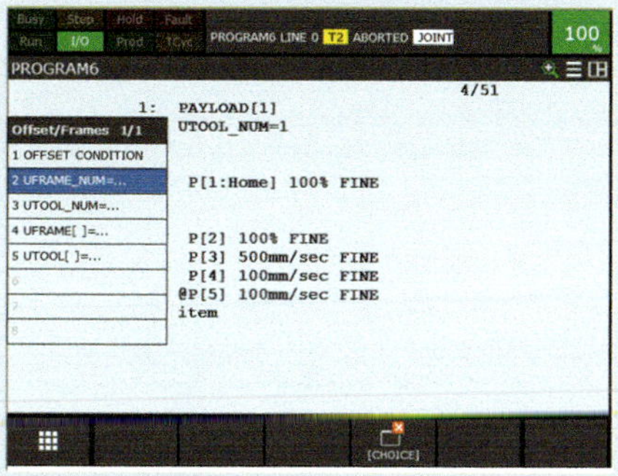

6. In the menu that appears, highlight Constant and press ENTER. The dots in the UFRAME_NUM=... are highlighted.
7. Using the teach pendant keypad, type the number of the desired user frame and press ENTER. In this case, the desired user frame is 0 (World).
8. If needed, repeat the procedure to add a user frame selection line before each movement line that uses a different user frame.

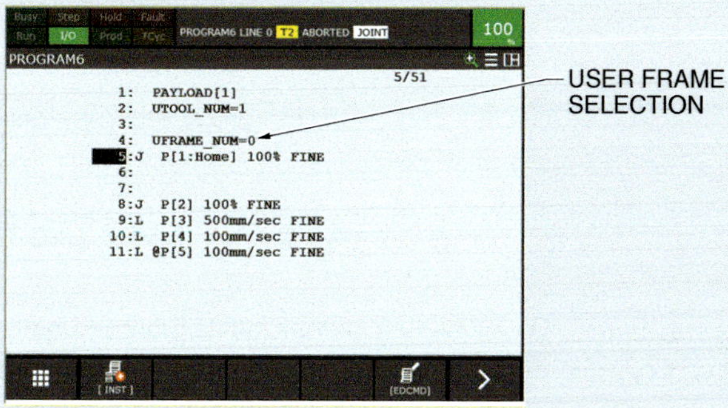

SECTION 6.2 USING THE PROGRAM ADJUST PROCEDURE

When new programs are first run, it is common to find robot movements that are very close to the desired action but not quite perfect. The programmer can make changes to the movement lines directly and redefine points, but there may be an easier alternative. **See Figure 6-2.** If the needed adjustments are small, the program adjust procedure can be used to automatically apply offset adjustments to lines of code. This is done while the program is running, which helps ensure that the adjustment is applied as needed. *Program adjust* is a procedure for making small adjustments to robot movements while it is running a program. Program adjust can change the robot's position (a maximum of 26 mm at a time), orientation, and speed. It can be used only for joint and linear movement types.

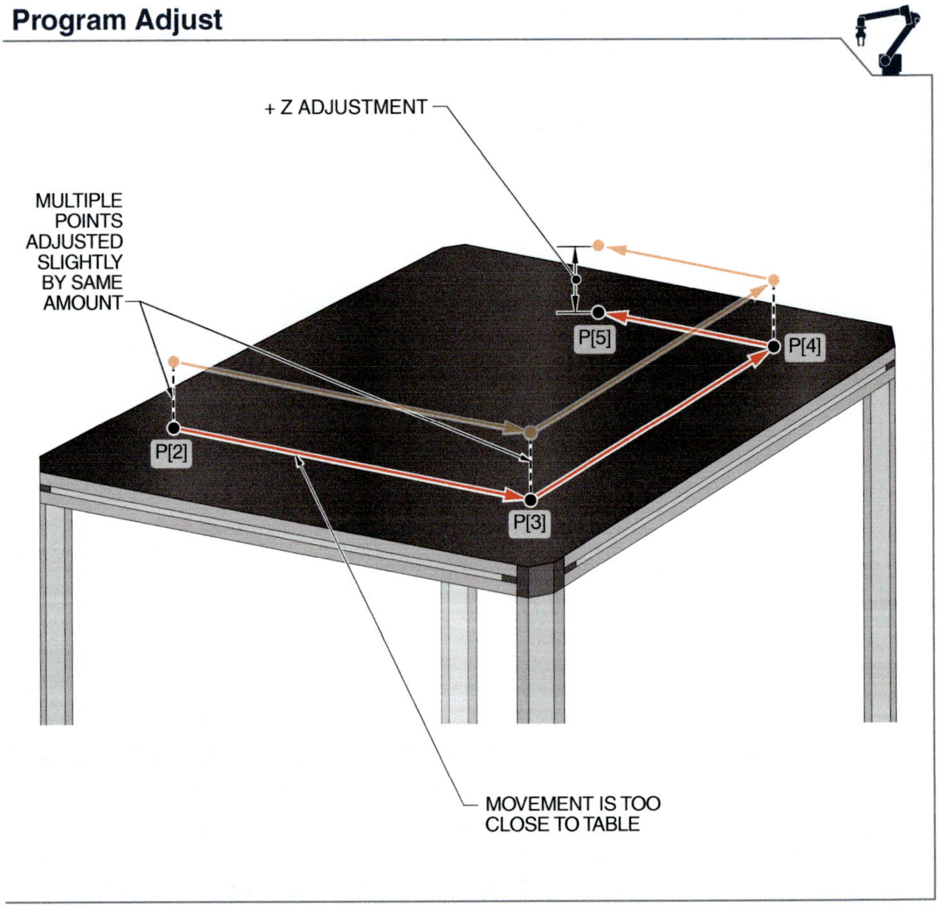

Figure 6-2. Program adjust is used to make small changes in movement line parameters for multiple points at once.

To use program adjust, the programmer first needs to make a program adjust schedule. **See Figure 6-3.** A *program adjust schedule* is a list of possible position, orientation, and speed offsets to be applied to one or more consecutive movement lines of program code. Then the program is run (in automatic mode) but at slow speed, and the adjustment(s) are applied. Once applied, the adjustments are permanent to the program. If the small adjustments are not sufficient, they can be applied multiple times to achieve the desired movement.

Program Adjust Schedule

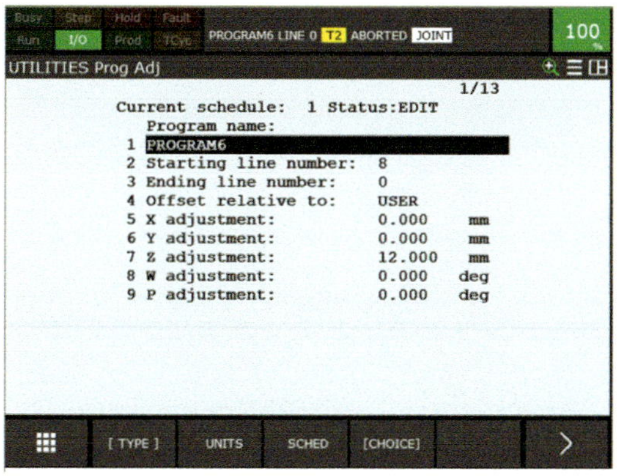

Figure 6-3. The program adjust schedule lists the incremental changes to position, orientation, and speed to be made to the movement lines.

As an example, a programmer has found that the tool travel was too close to a table. The table's surface is defined in a user frame. To move the tool in the +Z direction by 12 mm, the programmer creates a program adjust schedule to change four lines of code, lines 11 to 14. The effect of program adjust on these movement lines is that the positions of the points defined in them will be changed by 12 mm in the +Z direction. **See the *Using Program Adjust* Procedure.**

While not required for the procedure, it is helpful to switch to a double-screen display. The right side shows the program adjust schedule, and the left side shows the program to be adjusted. (Programs will only run while in the left-hand screen.)

PROCEDURE

Using Program Adjust

Identify Program Lines to Adjust

① Press SELECT.

② Select the desired program and press ENTER.

③ Move to the section of the program to be adjusted.

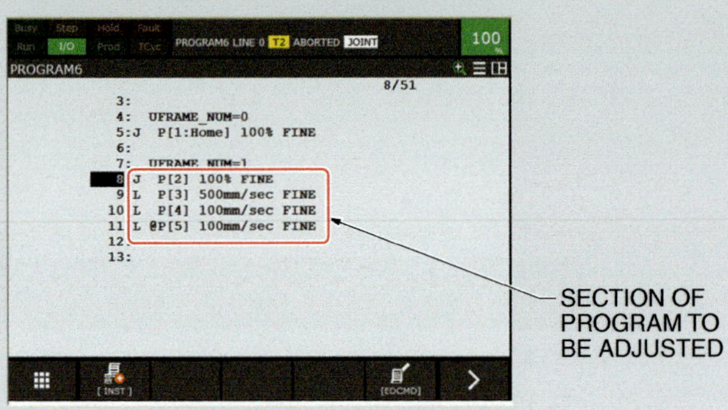

Set Up the Double-Screen Display

④ Press and hold SHIFT, and then press and release DISP to show the DISPLAY menu.

⑤ Highlight Double and press ENTER. The left screen will be used to show the program.

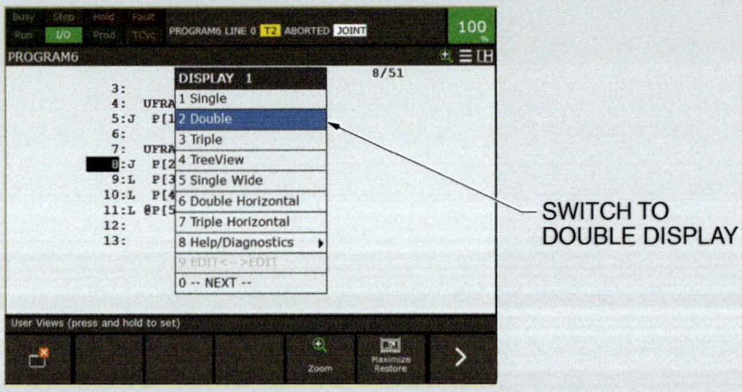

⑥ Touch the right screen to make it active. This will be changed to the program adjust schedule.

Make a Program Adjust Schedule

⑦ Press MENU.

⑧ Highlight UTILITIES and then Prog Adjust, and press ENTER.

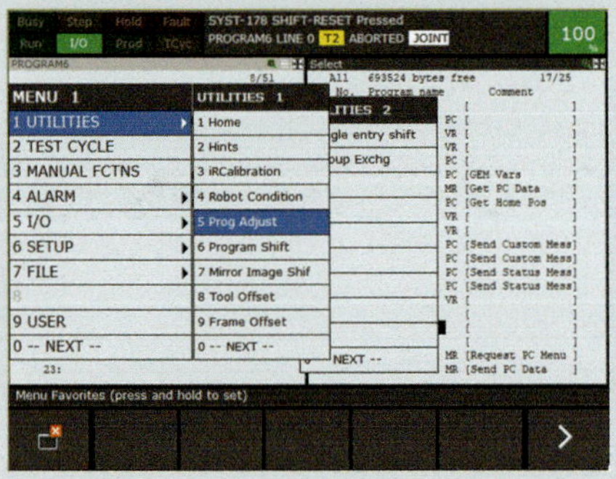

⑨ Highlight the desired schedule and press F2 `Detail`.

⑩ Verify that line 1 of the schedule shows the program to be adjusted. If the correct program is not shown, press F4 `Choice` and select the desired program.

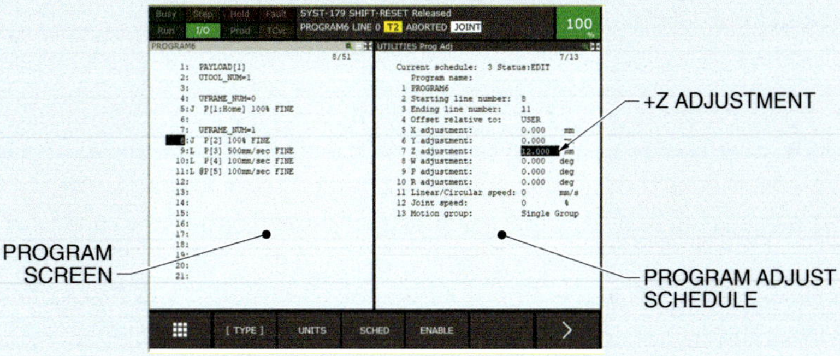

⑪ On line 2 of the schedule, enter the program line number where the adjustments are to start, and press ENTER.

⑫ On line 3 of the schedule, enter the program line number where the last adjustment is to be made and press ENTER. If only one program line needs adjustment, schedule lines 2 and 3 will contain the same line number.

⑬ On line 4 of the schedule, choose whether the offset will be relative to the current `USER` frame or `TOOL` frame. In this example, the adjustment is relative to the `USER` frame.

⑭ The only adjustment needed is in the Z direction, so on line 7 of the schedule, highlight the `0.00` and type 12 and press ENTER. No other lines require changes in this example.

Enable the Program Adjust Schedule

⑮ Touch to highlight the left (program) screen.

⑯ Using the speed ▽ button, reduce the speed to 10%.

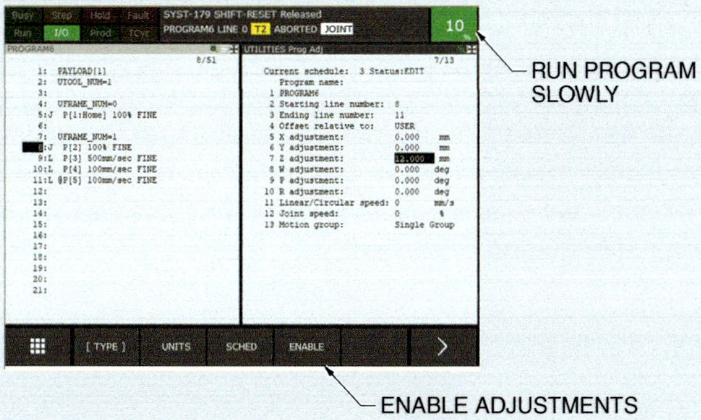

⑰ Run the program in Auto.

⑱ Highlight the right screen while the program is running.

⑲ Press F4 `ENABLE` to make the adjustments.

SECTION 6.3 DIRECTING PROGRAM EXECUTION

The lines of a program do not have to be executed in order. There may be reasons to have a program repeat sections of code, insert code from other files, jump to other locations in the program based on a conditional statement, or include other similar programming techniques. Two of the most common ways to direct program execution is with program calls and jump labels.

Program Calls

An industrial robot program can become extremely large and difficult to both understand and troubleshoot if the programmer has not properly planned and organized it. To prevent this, programs can be broken down into smaller subprograms that execute a defined section of the whole program. A *subprogram* is a program file with a section of code intended to be called by other programs for a specific task. **See Figure 6-4.** Programmers may also call these subroutines. The subprograms are created and edited in the same way as main programs. FANUC allows both TP programs and macros to be used as subprograms.

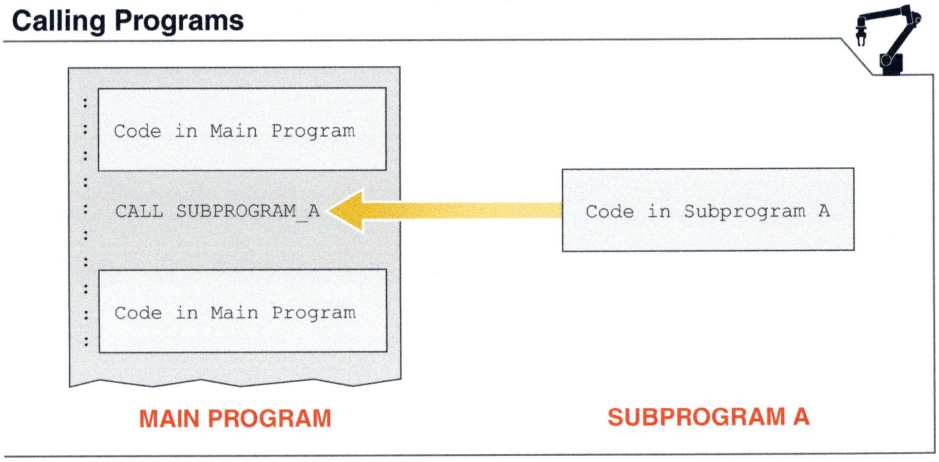

Figure 6-4. When calling a subprogram, which is a separate file, the result is the same as if that code was within the main program.

Subprograms are typically given names for a piece of equipment the robot is interacting with or a task the robot is executing. For example, an assembly program may have subprograms with names such as PlacePart1, PlacePart2, ApplyGlue, RotateAssbly, or PackBox.

The subprograms are then organized within a main program with Call instructions. When the robot reaches a program line with a Call instruction, it begins executing the specified subprogram. When that subprogram is completed, the robot returns to the main program and continues where it left off with the main program's next line. **See the *Calling a Subprogram* Procedure.**

PROCEDURE

Calling a Subprogram

1. In the main program, highlight the program line where the desired program is to start.
2. Press F1 [INST].

③ Highlight `CALL` and press [ENTER] to show the `CALL statement` menu.

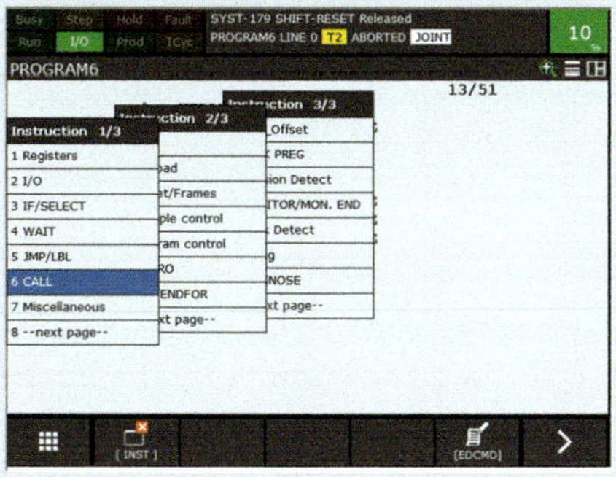

④ Highlight `CALL program` and press [ENTER] to show a listing of all available TP and macro programs.

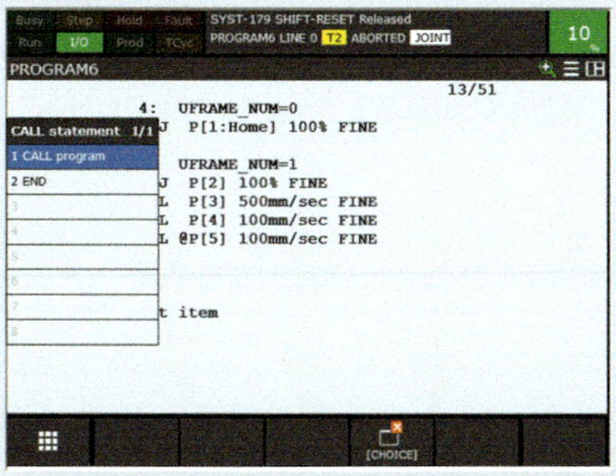

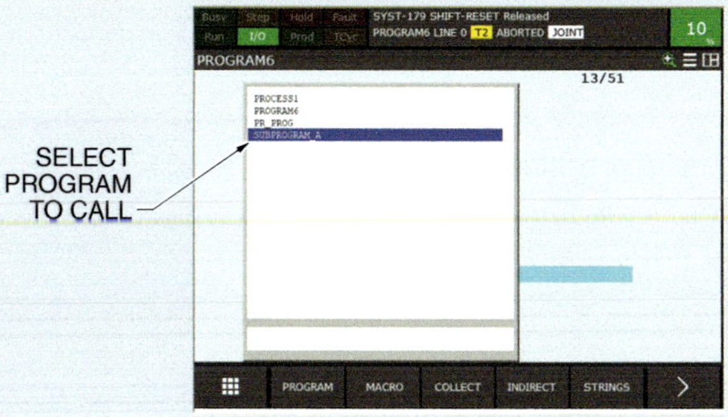

⑤ Highlight the desired program and press [ENTER].

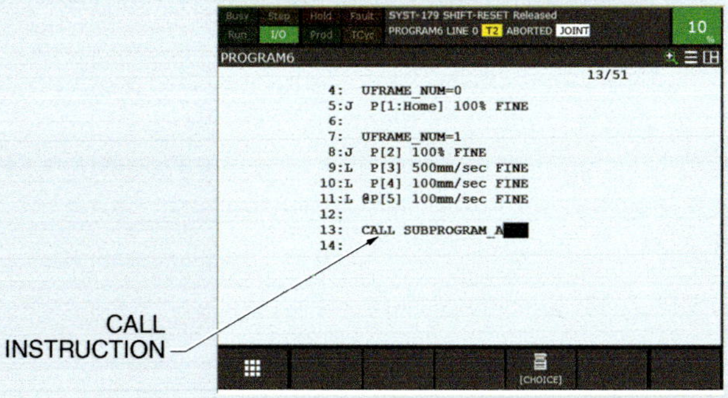

Very simple programs may not need to use Call instructions, but complex programs really benefit from this type of organization. Programs utilizing Call instructions are typically easier to read and understand, and they make it easier to locate certain parts of the main program. If the program must later be modified or a technician must troubleshoot the code, these well-defined subprograms help the programmer access the right code quickly.

Subprograms can even call other subprograms. This can be really useful but can also create problems if used incorrectly. Previously, in the Program Detail screen, the stack size was set to 500. This means that if a program calls another program, which calls another program, and so on, and this sequence of nested programs reaches 500 levels, the robot will show a Stack Overflow fault. It may seem like planned program calls could never reach this large number. However, this situation can easily happen by accident. If Program A calls Program B, and then Program B calls Program A, a program loop is created where the calls never end, and the stack overflow is quickly reached. **See Figure 6-5.** To avoid this problem, Program B should be allowed to run to its end, and it will return to Program A automatically.

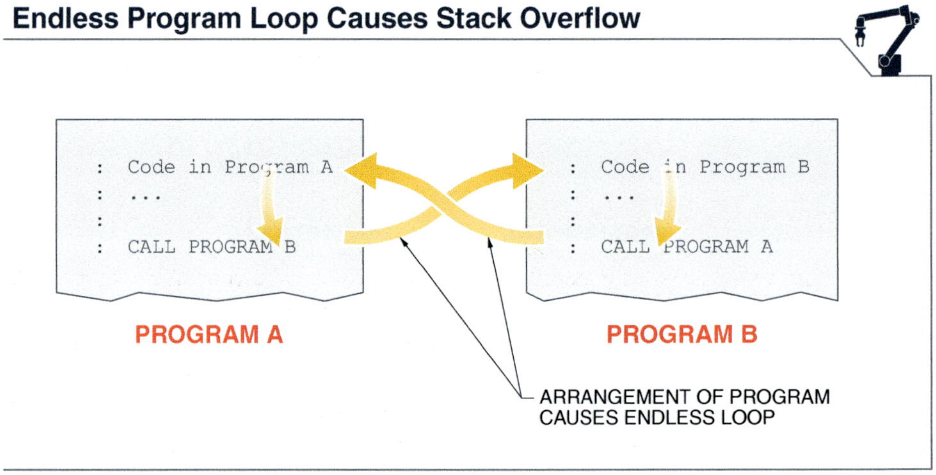

Figure 6-5. Calling a program incorrectly can result in an endless loop of program execution. This can cause a Stack Overflow fault.

Jump Labels

Robots are particularly suited for repetitive tasks, repeating the same series of steps many times. An easy way to program repeating sequences of code is to use jump labels. A *jump label* is a set of two instructions that directs the execution of code from one part in a program to another part of the same program. The program execution "jumps" from one point to another, which is identified by a label.

This can be used to repeat a section of code. The programmer only needs to create the section once and then use the jump label instruction to make the program repeat the desired code many times. **See Figure 6-6.**

Jump Labels

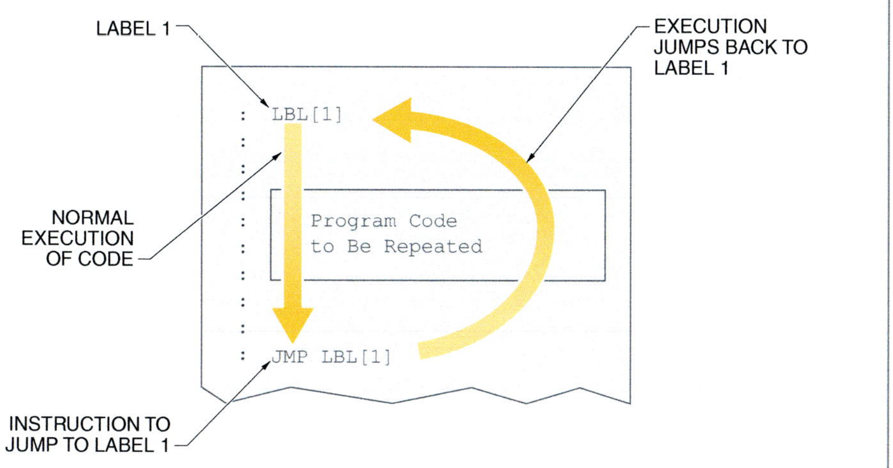

Figure 6-6. A jump label is used to repeat sections of code by "jumping" from one line to another.

The programmer must first decide which lines need to be repeated. On a blank line, just before the first line to be repeated, the programmer places the label instruction `LBL[...]`. Between the brackets, the programmer enters a unique number between 1 and 32,766. This is an identifying number, not the number of times the robot should repeat these movements. Then the instruction `JMP LBL[...]` is placed on a blank line directly after the last movement line to be repeated. The number between these brackets is the same as the one entered on the LBL line. Then the `LBL[...]` becomes the destination of the `JMP LBL[...]` of the same number. **See the *Adding a Jump Label Instruction* Procedure.** Using a jump label alone creates an endless loop of code. Other programming techniques, covered in Chapter 7, are added to such codes to control how many times sections are repeated.

There are a few rules when using the jump label instruction:

- The instructions `JMP LBL[...]` and `LBL[...]` must be in the same program. For example, `JMP LBL[6]` cannot be in one program and refer to a `LBL[6]` in another program.
- There can be only one `LBL[...]` destination for each number per program. If the programmer assigned `LBL[8]` in the program already, `LBL[8]` cannot be used again.
- There may be multiple `JMP LBL[...]` instructions to the same label. For example, all `JMP LBL[8]` instructions have the same destination of `LBL[8]`.

PROCEDURE

Adding a Jump Label Instruction

Add the Label

① Highlight a blank line prior to the lines to be repeated.

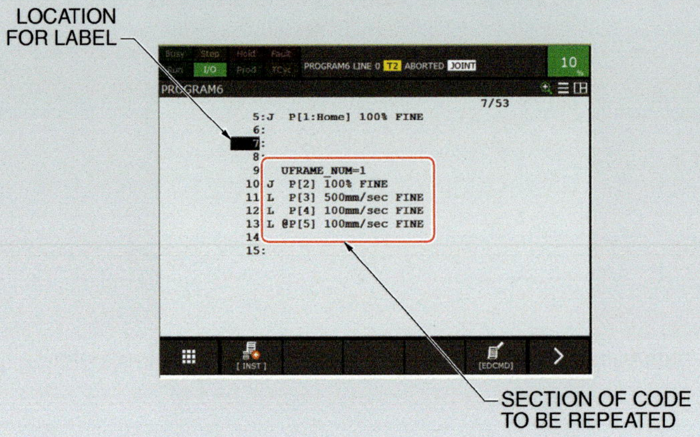

② Press F1 `[INST]`.

③ Highlight `JMP/LBL` and press ENTER.

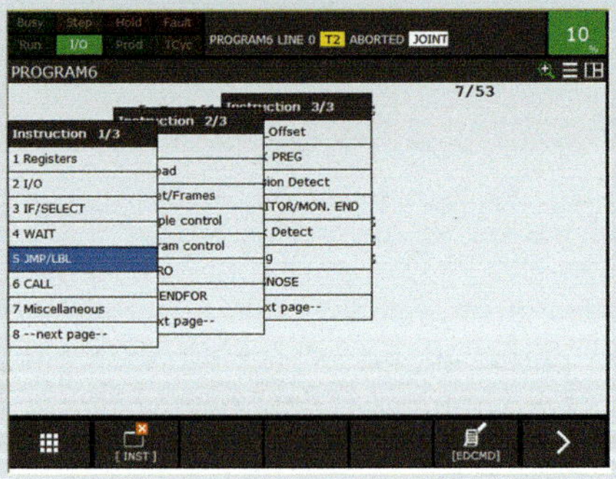

④ Highlight `LBL` [...] and press ENTER.

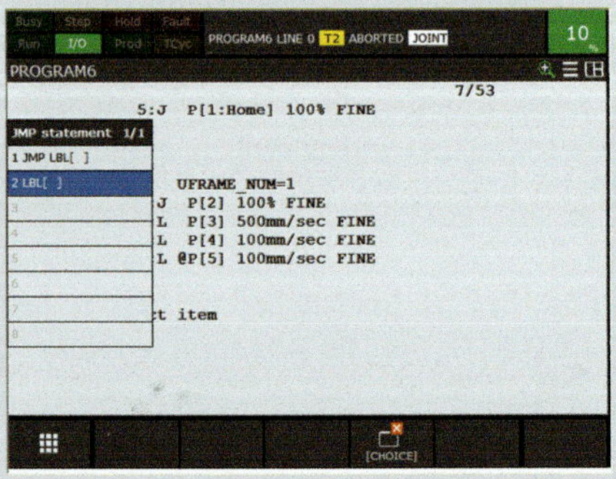

⑤ The space between the brackets is highlighted. Use the teach pendant keypad to assign a number between 1 and 32,766 and press ENTER.

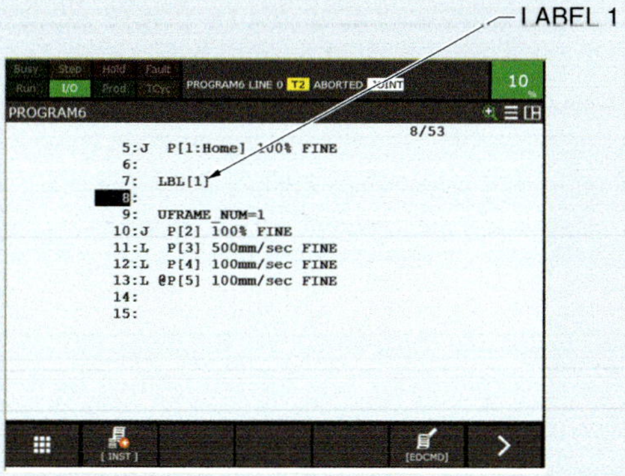

Add the Jump Label Instruction

⑥ Highlight a blank line below the last movement line to be repeated.

⑦ Press F1 `[INST]`.

⑧ Highlight `[INST]` and press ENTER.

⑨ Highlight `JMP LBL[...]` and press ENTER.

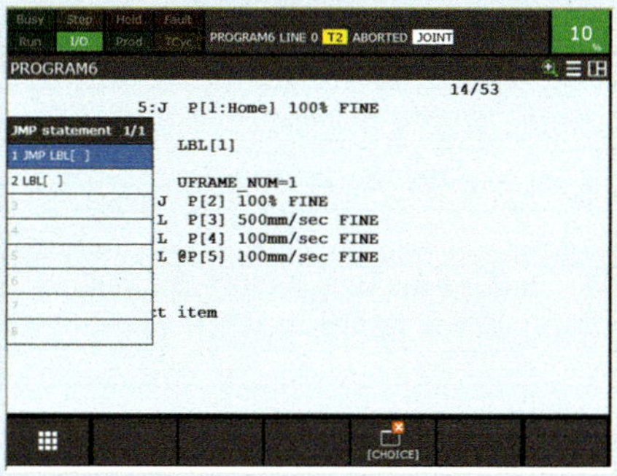

⑩ The space between the brackets is highlighted. Use the teach pendant keypad to input the same number as the destination label.

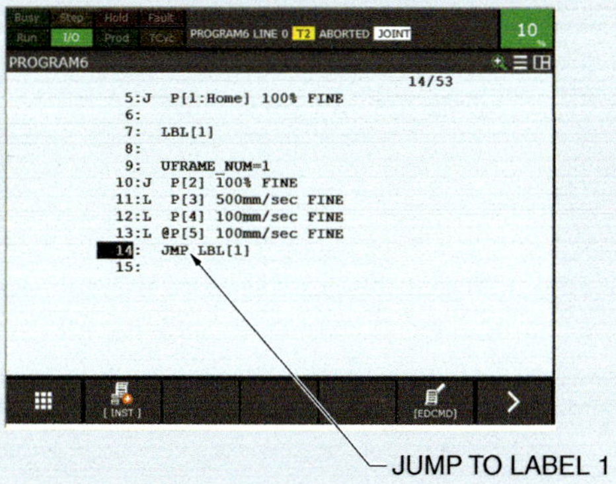

SECTION 6.4 FINE-TUNING PROGRAM TIMING

Much of a programmer's time is spent fine-tuning a program to exactly match the timing of the robot to production requirements and to synchronize the robot with other processes it interacts with. For example, if a robot that performs one process in a production line is slower than the other processes, product will back up and cause the entire line to slow down. Carefully managing program timing can improve efficiency for the whole line. Programmers can use timers to measure the execution time of sections of code and the Wait instruction to selectively delay at certain moments.

Timers

Robot programs are typically optimized for minimizing cycle time. *Cycle time* is the time it takes a robot to complete one set of programmed actions, which is then repeated. In automated production processes, cycle time is frequently measured and optimized to within tenths or hundredths of a second.

In the planning stages, engineers use calculations and simulations to arrive at estimated cycle times for various production options. When a robot is chosen for a process, this estimated cycle time becomes the programmer's goal. The programmer tests and adjusts the program with the actual robot and real product in order to meet the required cycle time. If that cycle time cannot be reached, the robot installation may be decommissioned, costing the company thousands of dollars in capital, wasted resources, and lost production. Examples of programming decisions that affect cycle time include the following:

- Robot Speed–A faster speed does not always mean the robot moves more efficiently.
- Movement Type–Generally, a joint movement is faster than a linear movement.
- Termination Type–CNT is always faster than Fine. However, CNT50 may be faster than CNT60.
- Robot Positioning–The robot can wait close to the part or in the Home position. For example, it can wait in the Home position while another machine completes a task.

Programmers may test these and other variations to identify the fastest move without sacrificing personnel safety or product quality.

Cycle time is measured with timers. A *timer* is a programming instruction that measures the robot's physical execution time of a section of code. **See Figure 6-7.** The effects of various programming changes on cycle times are measured and compared to determine the best choices. Typically, a program is divided into logical sections to be timed individually. By breaking the program into small sections and making changes to each section one at a time, the programmer can more easily identify which changes result in an improved cycle time and which do not.

Instruction lines are added to a program that start, stop, and reset each timer, similar to a stopwatch. **See the *Adding a Timer* Procedure.** A timer can time a part of a program, a whole program, or even multiple programs. A timer can be started in one program and stopped in another. They can also be started and stopped multiple times before resetting. Multiple timers can be used within one program and in any order. The FANUC robot software allows for 20 timers. Each timer has a comment section. However, the comment is not displayed when the timer is entered into the program.

Since timers can span programs and can be started and stopped multiple times, they must be reset in order to return to zero. Resetting a timer should always be done at the beginning of a program. If the reset were done at the end of the program, the programmer would never have a chance to read the results.

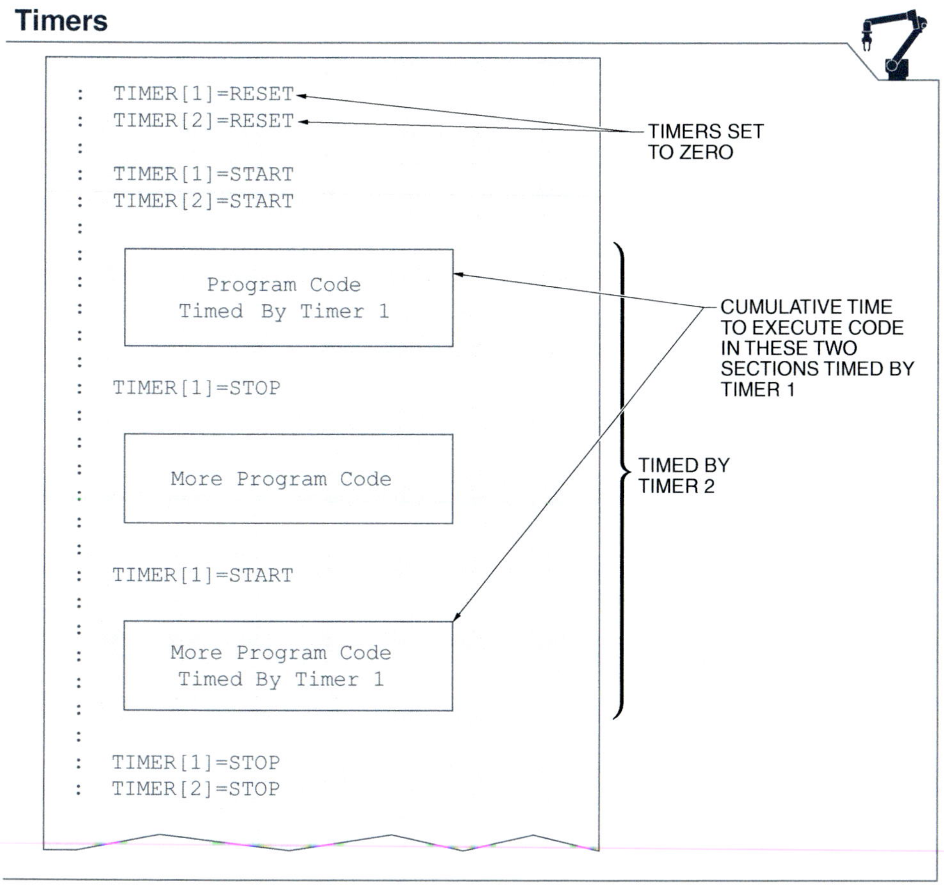

Figure 6-7. A set of timer instructions encloses the section of code to be timed. It is important to reset the timer to zero before starting a new timing.

A timer is started in the line prior to the section to be timed and stopped in the line directly after the section. For example, if timing a circular movement, the timer instructions are placed right before and after the circle so that it is only tracking the time it takes to complete the circle. This would not include the time for the robot to move from the Home position to the start point of the circle or from the end of the circle back to the Home position.

PROCEDURE

Adding a Timer

Name the Timer

1. Press [MENU].
2. Highlight `NEXT` and press [ENTER].
3. Highlight `STATUS` and then `Prg Timer`. Then press [ENTER]. The `PRG TIMER LISTING` screen is shown.

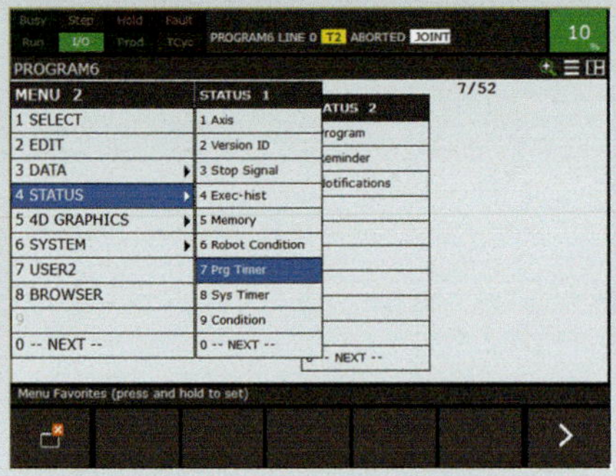

4. In the comment column, give `TIMER 1` a descriptive name by pressing [ENTER] and using the keyboard to type a name.

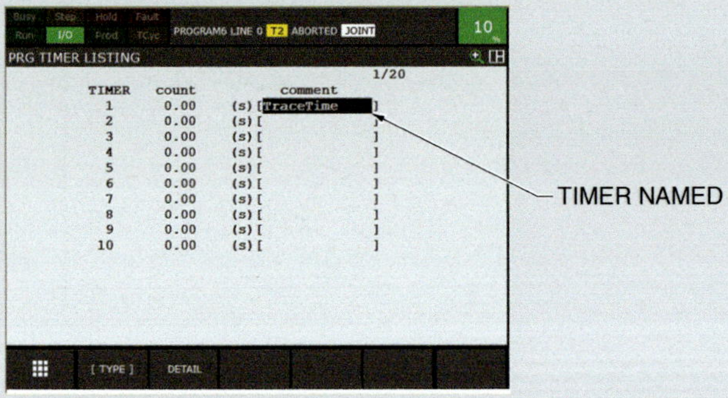

Insert the Reset Timer Instruction

5. Open a program by pressing [SELECT], highlighting the desired program, and pressing [ENTER].
6. Highlight a program line near the top of the program.
7. Press [F1] `[INST]` to show the `Instruction` menu.
8. Highlight `Miscellaneous` and press [ENTER].
9. Highlight `TIMER[ ]` and press [ENTER].

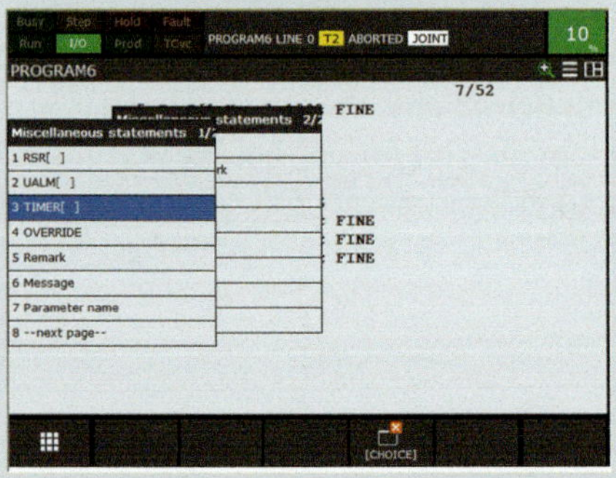

10. The area in the timer instruction brackets is highlighted. Type the desired timer number (1 through 20) using the number pad on the teach pendant. If the desired number is not known, press [F5] `[LIST]` to show the list of timers.

⑪ Once the number of the desired timer is entered, a menu appears. Highlight `RESET` and press [ENTER].

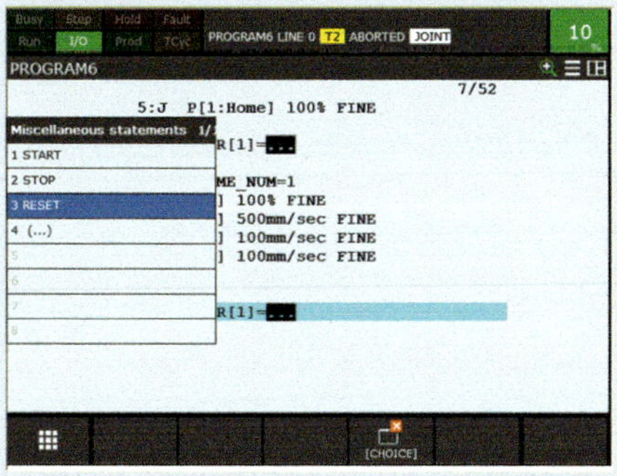

Insert the Start Timer Instruction

⑫ Highlight the program line just before the section of code to be timed. If there is not an open line, insert a line.

⑬ Press [F1] `[INST]` to show the `Instruction` menu.

⑭ Highlight `Miscellaneous` and press [ENTER].

⑮ Highlight `TIMER[ ]` and press [ENTER].

⑯ Type the same timer number used in the reset instruction.

⑰ From the menu, highlight `START` and press [ENTER].

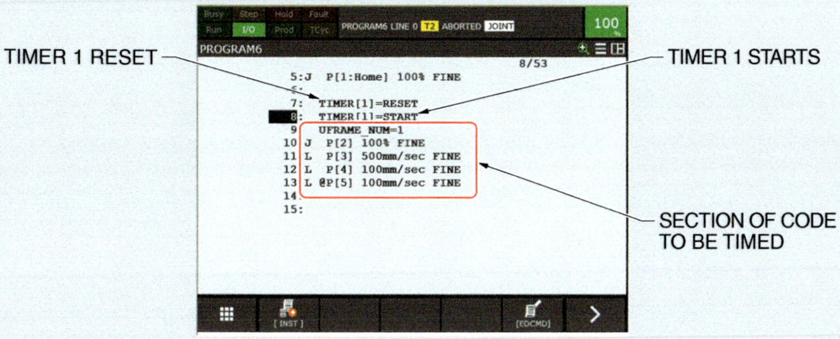

Insert the Stop Timer Instruction

⑱ Highlight the program line just after the section of code to be timed. If there is not an open line, insert a line.

⑲ Press [F1] `[INST]` to show the `Instruction` menu.

⑳ Highlight `Miscellaneous` and press [ENTER].

㉑ Highlight `TIMER[ ]` and press [ENTER].

㉒ Type the same timer number used in the start instruction.

㉓ From the menu, highlight `STOP` and press [ENTER].

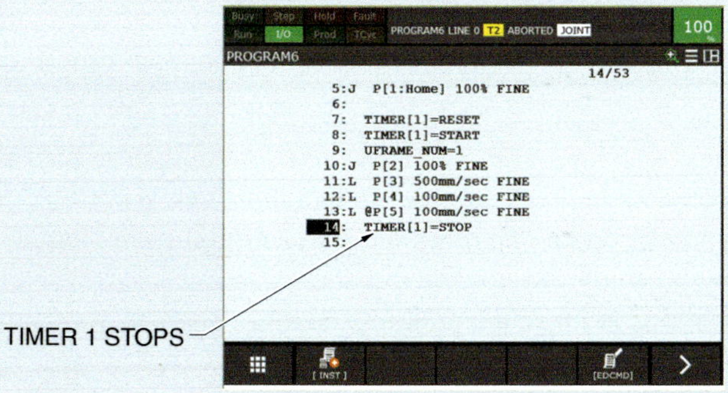

As the program runs, the elapsed time can be viewed on the teach pendant, but it is displayed on a separate screen, not in the program editing screen. **See the *Reading a Timer* Procedure.**

PROCEDURE

Reading a Timer

1. Press MENU.
2. Highlight -- `NEXT` -- and press ENTER.
3. Highlight `STATUS` and then `Prg Timer`. Press ENTER. The `PRG TIMER LISTING` screen is shown.
4. The elapsed time is shown in the count column. More information on the timer can be seen by pressing F2 `DETAIL`.

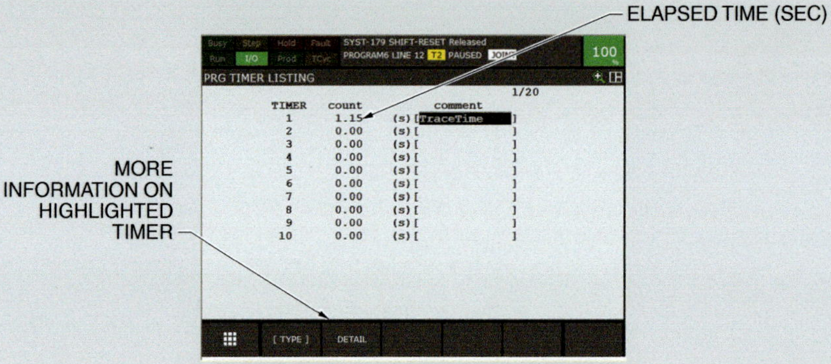

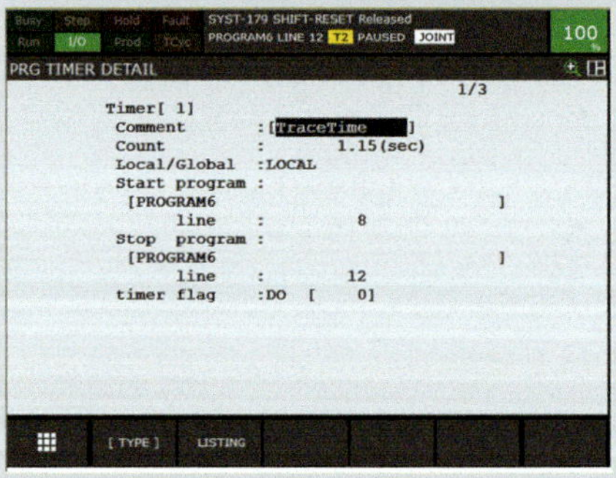

Wait Instructions

The Wait instruction holds the robot in its position for a certain amount of time. Wait instructions have two parts. The first is `Wait`, and the second part describes what the robot is waiting for.

This sounds simple, but this instruction is versatile. The amount of time the robot waits can be based on things such as operator inputs, the conditions in the cell, or the status of another machine. For example, an industrial robot may need to interact with other machines or devices, which requires some sort of communication between them in order to coordinate their actions. For example, a robot must load a product through a door but only when the door is open. The robot receives a digital signal letting it know if the door is open. A digital signal is on when the door is open and off when the door is closed. Just prior to the code to move the product, the robot's program can use a Wait instruction, which will pause the program and wait for the on signal, meaning that the door is open. When the on signal is received, the robot continues the program.

Working with additional inputs and variables, such as the ones in this application, is covered in a later chapter. However, Wait can also be used as a simple instruction to pause the robot for a defined length of time in seconds. This is simple and still very useful. **See the *Adding a Wait Instruction* Procedure.**

PROCEDURE

Adding a Wait Instruction

① Highlight a blank line.

② Press F1 [INST].

③ Highlight WAIT. Press ENTER to show the Wait statements menu.

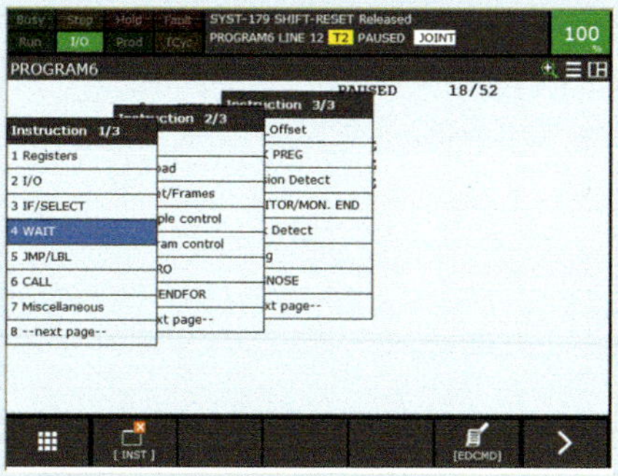

④ Highlight WAIT...(sec) and press ENTER.

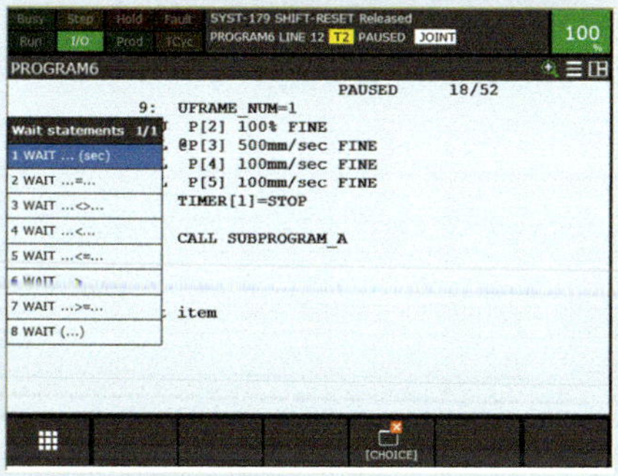

⑤ The 0.00 prior to the word (sec) will be highlighted. Enter the number of seconds desired for the robot to wait.

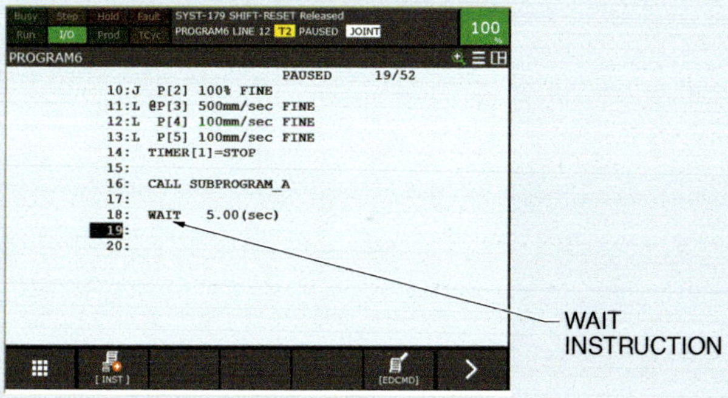

SECTION 6.5 ADDING PROGRAM NOTES

Remarks and messages are simply readable notes contained in the program. However, remarks are intended for programmers, and messages are intended for operators.

Remarks

A *Remark instruction* is an added instruction within program code that has no effect on the robot and is intended as a plain-language note by a programmer to explain a section of code. In most cases, remarks are used to assist other programmers or maintenance personnel in understanding what the next lines of code are for. For example, a programmer may want to document that tracing a green table is the first step in production. The remark entered is, "First step in production GrnTbl." A remark can be up to 32 characters in length. **See the *Adding a Remark* Procedure.**

The Remark instruction is different from the Remark command. That command is used to temporarily ignore lines of existing code.

PROCEDURE

Adding a Remark

1. Highlight the desired blank line. (If needed, use the `[EDCMD]` menu to insert a blank line.)
2. Press F1 `[INST]`.
3. Highlight `Miscellaneous` and press ENTER.
4. Highlight `Remark` and press ENTER.

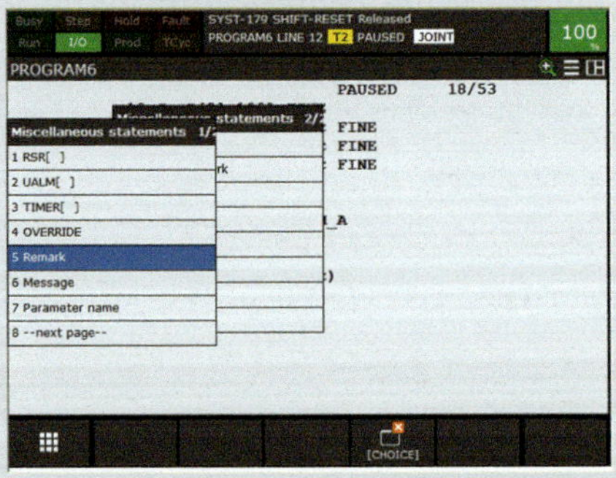

5. Press ENTER again to show the `Alpha Input 1` menu.
6. Highlight `Options/Keybd` and press ENTER.
7. Type the desired remark and press EXIT.

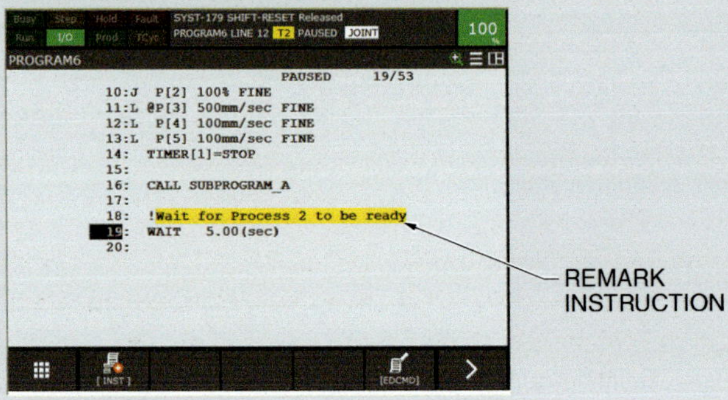

Messages

A *message* is a note that appears (typically on the teach pendant) for the operator when the robot reaches a certain point in its program. A message can be up to 32 characters in length. Messages are often used to inform the operator that a robot is waiting for them to complete a task, such as a part-quality inspection and/or to provide an input (such as a button press) to the robot. If the operator's quality inspection were to find a substandard part, the robot's pause allows changes to be made before it continues producing more substandard parts. **See the *Adding a Message* Procedure.**

PROCEDURE

Adding a Message

1. Highlight the desired blank line.
2. Press F1 [INST].
3. Highlight Miscellaneous and press ENTER.
4. Highlight Message and press ENTER. The instruction Message [...] appears with the space between the brackets highlighted.

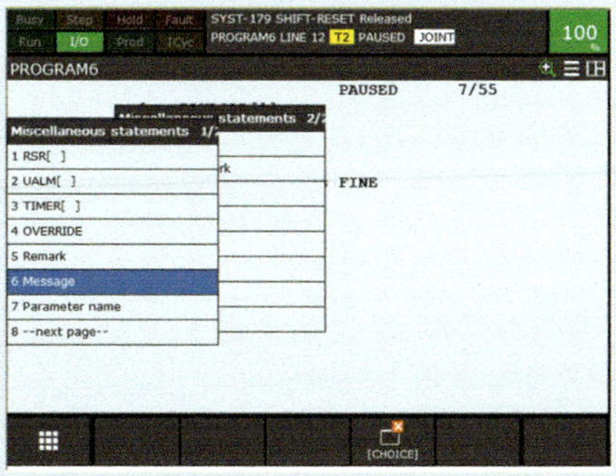

5. Press ENTER again to show the Alpha Input 1 menu.
6. Highlight Options/Keybd and press ENTER.
7. Type the desired message and press EXIT.

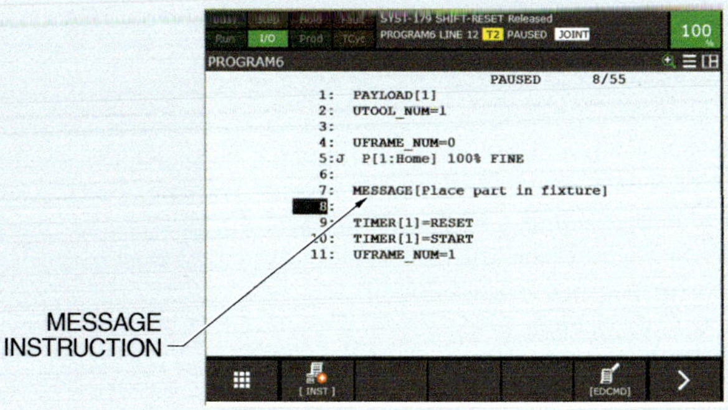

The message appears on the user screen. The *user screen* is the display used for showing operator messages. **See Figure 6-8.** When using the teach pendant, the display automatically changes to the user screen whenever a message is encountered in the program code. When a robot is set up for production work and the teach pendant is not regularly used, a separate display can be set up to show messages.

User Screen

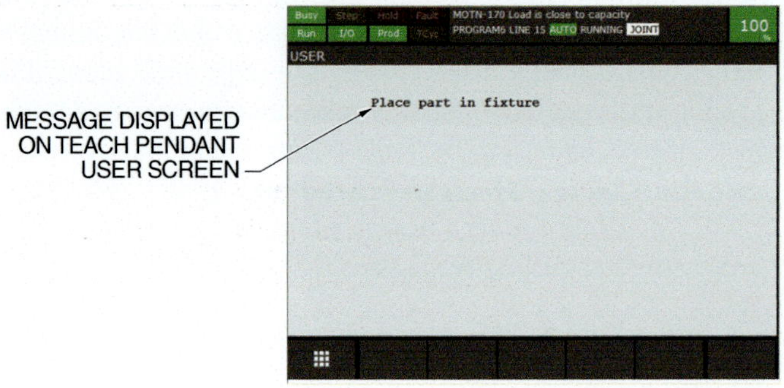

Figure 6-8. Messages are displayed on the user screen. These messages are meant for the operator.

Review 6

Name ______________________________ Date ______________

SECTION 6.1 SELECTING FRAMES

______________ **1.** Prior to creating a movement line, the programmer decides which tool and user frames to use and selects them in the ___ menu.

A. Instruction
B. Jog
C. Frame
D. Initialization

______________ **2.** When a user frame selection instruction is not included when needed, the result is a(n) ___ fault.

______________ **3.** If only one tool or user frame is used for an entire program, the selection line is put ___ the program.

A. at the end of
B. on any blank line in
C. after the first movement line in
D. at the beginning of

4. When are instruction lines for selecting user frames and tool frames required?

SECTION 6.2 USING THE PROGRAM ADJUST PROCEDURE

______________ **5.** Program adjust can change the robot's position, orientation, and ___.

A. group
B. speed
C. tool frame
D. waiting time

______________ **6.** A program adjust ___ is a list of possible position, orientation, and speed offsets to be applied to one or more consecutive movement lines of program code.

______________ **7.** Once applied, program adjust offsets are ___.

A. permanent
B. temporary
C. deleted
D. unrepeatable

8. When a program adjust schedule is applied, how many consecutive movement lines can be changed?

9. Why is the program adjust procedure used while the program is running?

SECTION 6.3 DIRECTING PROGRAM EXECUTION

_______________ 10. A(n) ___ is a program file with a section of code intended to be called by other programs for a specific task.

_______________ 11. Subprograms are organized within a main program with ___ instructions.

A. Wait
B. Jump
C. Call
D. Message

_______________ 12. If Program A calls Program B, and then Program B calls Program A, a program loop is created where the programs never end and the ___ is quickly reached.

_______________ 13. A ___ is a set of two instructions that directs the execution of code from one point in a program to another part in the same program.

A. wait
B. jump label
C. program call
D. loop

14. How are subprograms typically named?

15. What happens when a called program is completed?

16. How are label locations identified?

17. Why are jump labels typically used?

18. Can there be multiple label instructions and jump label instructions within a program?

SECTION 6.4 FINE-TUNING PROGRAM TIMING

__________ **19.** ___ is the time it takes a robot to complete one set of programmed actions, which is then repeated.

__________ **20.** Resetting a timer should always be done at the ___ of a program.

A. beginning
B. middle
C. end
D. called subroutine

21. What four parameters can be adjusted in a program to affect cycle time?

22. How are timers used to improve cycle time?

23. What are the two parts of a wait instruction?

SECTION 6.5 ADDING PROGRAM NOTES

__________ **24.** A(n) ___ is a line of code where no action is passed to the robot.

A. remark
B. message
C. command
D. instruction

__________ **25.** Messages are often used to inform the ___ that a robot is waiting for them to complete a task, such as a part-quality inspection and/or to provide an input to the robot.

A. maintenance technician
B. programmer
C. operator
D. supervisor

26. Why are remarks commonly used?

6 Lab Exercise

Lab Objective

Add various instruction lines to enhance the code in an existing object-tracing program.

Equipment Required

- Robot with end effector, pointer, and associated tool frame
- "SquareTrace" program from Lab Exercise 4
- "Circle" macro from Lab Exercise 5
- Two square objects, placed in the robot cell, and their associated user frames

Lab Procedure

1. Select the tool frame on line 2.
2. Select the user frame for the first square in a line right before the trace of that square.
3. Add the instruction lines for a timer that times the tracing of the first square.
4. Add a remark prior to tracing the first square that says, "First step in production.
5. Call the "Circle" macro in a line between the two square traces.
6. Add a message prior to the macro call that says, "Second step is repeated."
7. Select the user frame for the second square in a line right before the trace of that square.
8. Place a label instruction (number "1") prior to the macro call.
9. Place a "jump to label" (JMP LBL) instruction after the second square trace and use the same label number.
10. Add the instruction to wait for 2 sec before repeating the trace of the second square.

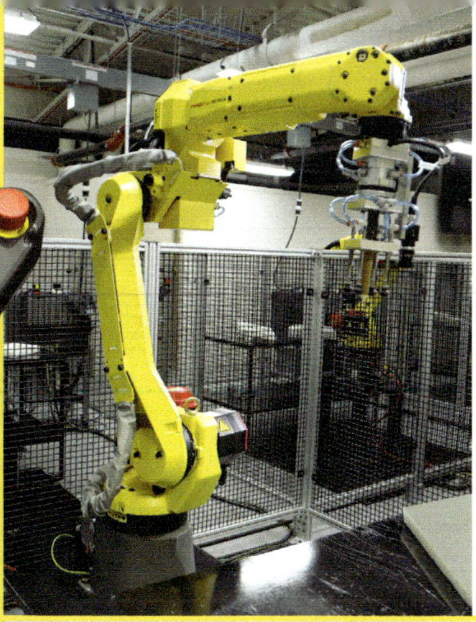

7
Registers

Robot programs are commonly adapted for repeating processes a certain number of times, producing varying numbers of products, and retooling for different models of products. Manufacturing demands often require that robots and programs be somewhat flexible to optimize production. One programming tool used to do this is the use of registers. These placeholders can be used to control operations, count products, or adapt to different processes. A common application of registers is to repeat a section of code a certain number of times.

Objectives

SECTION 7.1 UTILIZING REGISTERS

- Describe how to set up a register for use in a program.
- Demonstrate the use of the three command lines used in a register-based counting program.

SECTION 7.2 UTILIZING POSITION REGISTERS

- Differentiate between a position and a position register.
- Demonstrate how to change a position to a position register.
- Describe two useful features of position registers.

Key Terms

- register
- position register (PR)

Learner Resources

atplearningresources.com
Access Code: **762316**

SECTION 7.1 UTILIZING REGISTERS

A *register* is a software variable used to store a number. There are five types of registers, but two are most commonly used—registers (general-purpose) and position registers. There are 200 available, general-purpose registers, which can hold any number from 0 to many thousands or millions depending on the controller model. Registers are identified with a short name in the DATA Registers screen. This name appears next to its number every time it is used in a program.

Command lines in a program are used to set the value of a register, change the value with a mathematical statement, or use the value to affect some part of the program. There are many ways in which registers are useful, but a common application is using them to count the number of times a section of code is repeated before the robot moves on to the rest of the program. Jump labels can be used to repeat a section of code, but they have no way of limiting the number of repeats, so they will repeat the section forever. By using a register with the jump label, each execution can be counted. **See Figure 7-1.** Then another command line (IF statement) can end the looping when the desired number is reached.

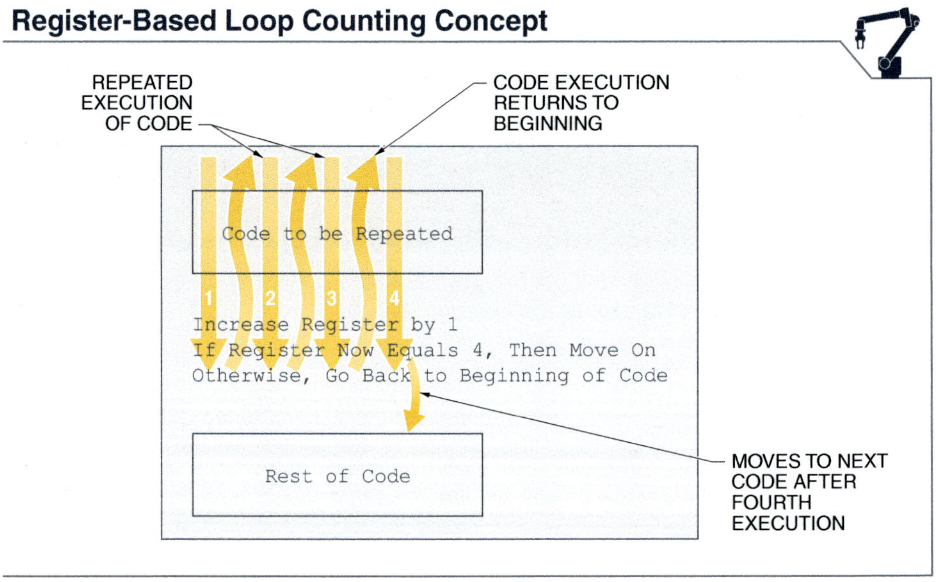

Figure 7-1. A register holds the number of times a loop has repeated so that this information can be used to direct the path of execution.

For example, consider an industrial application in which a robot must complete two processes on the same product. Process 1 is completed four times on the product before the robot moves on to Process 2. Process 2 is completed once.

After a register is named, it is typical to set its initial value (typically 0) at the beginning of a program. It must be set to zero before the repeated code, so it is usually placed at the beginning of the program. A jump label is used to repeat Process 1. In order to count each time Process 1 is completed, a statement at the end of the process (but before the JMP LBL command) increases the value of the register by 1. **See Figure 7-2.** As Process 1 repeats, the register increases by 1 each time. Then a conditional statement declares that when the count equals 4, instead of continuing to the next line and looping again, the robot jumps to a new label. This label location is defined after the original jump loop and allows the robot to continue on to Process 2. **See the *Using a Register to Count Code Loops* Procedure.**

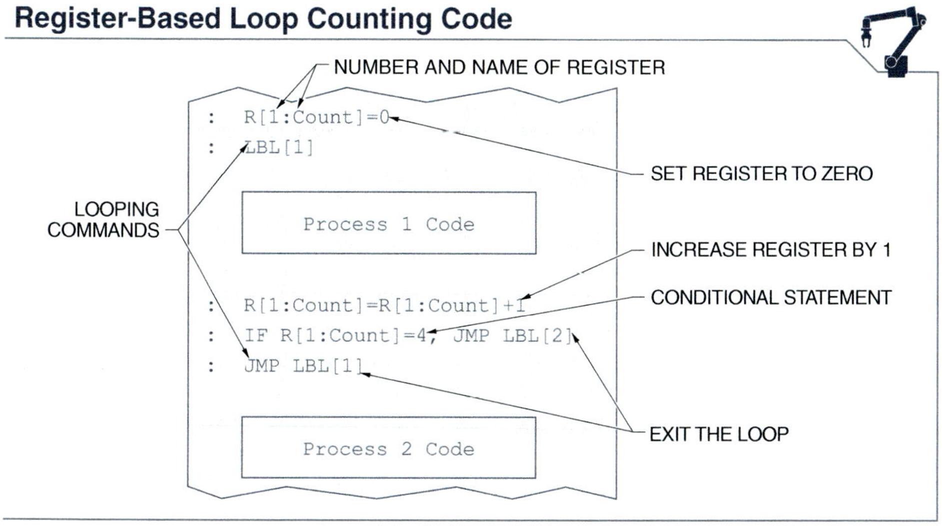

Figure 7-2. A conditional statement compares the count register to the desired value and either lets the loop repeat again or exits the loop and directs execution elsewhere.

To take the example further, consider that the robot in the example must produce six units of this product and then stop. Another jump label loop and a second register can be used to enclose both processes and allow it to repeat until the count reaches six and then stop. **See Figure 7-3.** Such code structures can be nested and sequenced in many different ways to create sophisticated programs.

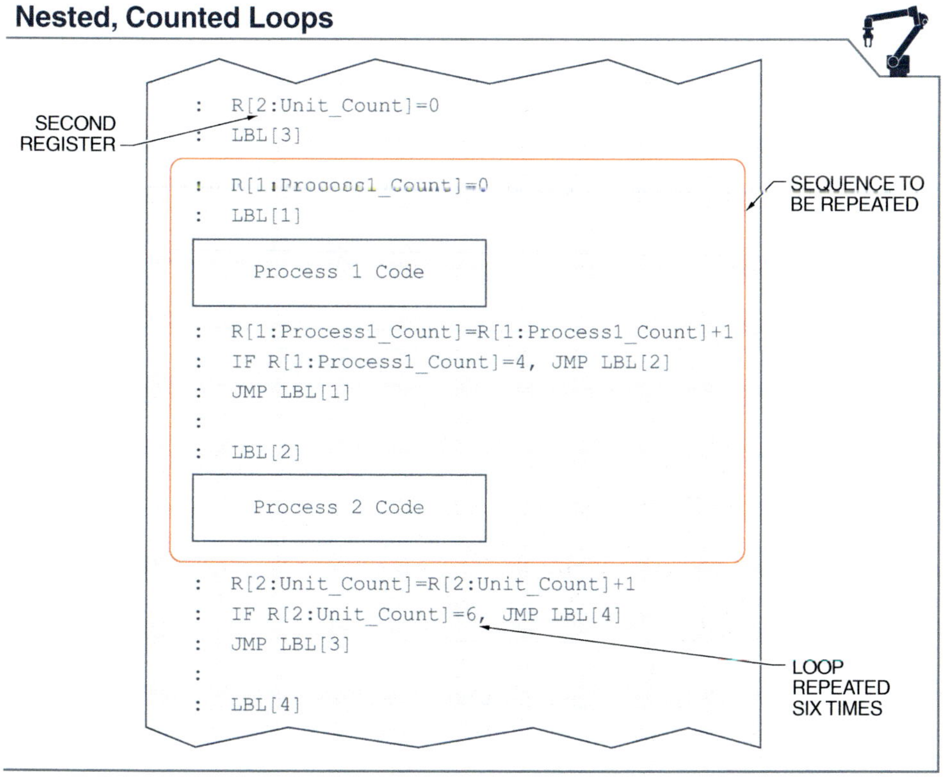

Figure 7-3. Counted loops can be nested or sequenced to create complex programs.

PROCEDURE

Using a Register to Count Code Loops

Set Up the Register

① Press DATA.

② Press F1 [Type].

③ Highlight Registers and press ENTER to show the DATA Registers screen.

④ Highlight the desired register and press ENTER to add a comment identifying the register.

⑤ Highlight Options/Keybd, press F5 Keyboard, and type a comment.

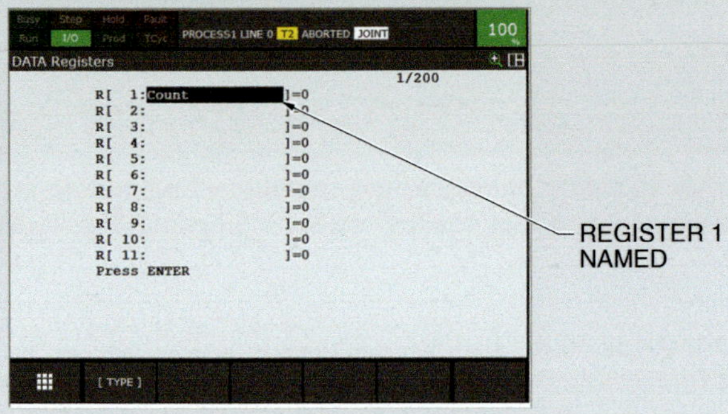

Set the Process Count Register to Zero

⑥ Open the desired program by pressing SELECT, highlight the program name, and then press ENTER.

⑦ Highlight a blank line directly above the command line LBL used for repeating the first process.

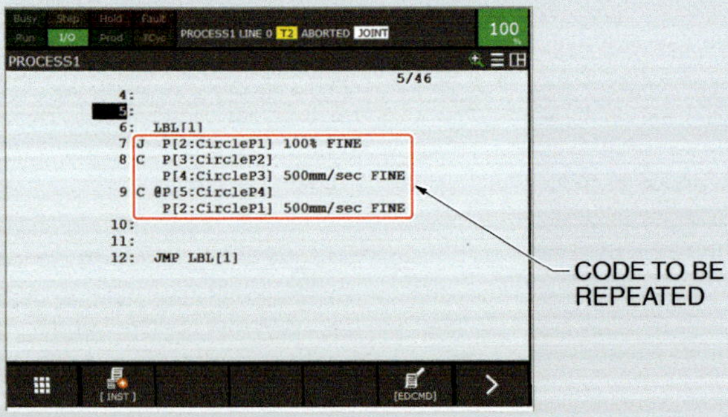

⑧ Press F1 [INST] to show the Instruction menu.

⑨ Highlight Registers and then press ENTER to show the Register Statement menu.

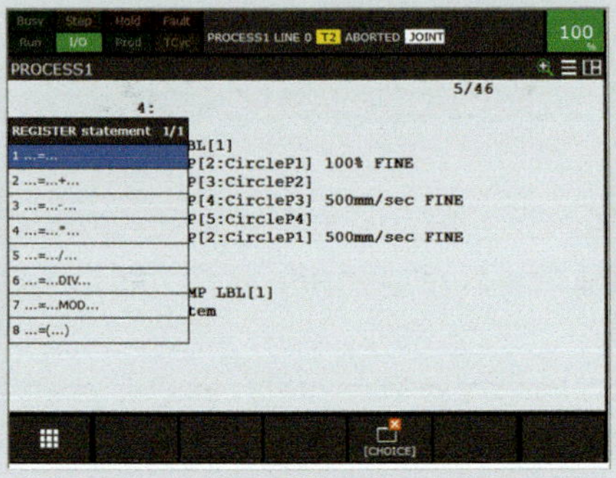

⑩ Highlight ...=... and press ENTER. The text ...=... appears in the line. The first set of dots is highlighted, and the `Register Statement` menu appears.

⑪ Highlight `R[ ]` and press ENTER. The space between the brackets is highlighted.

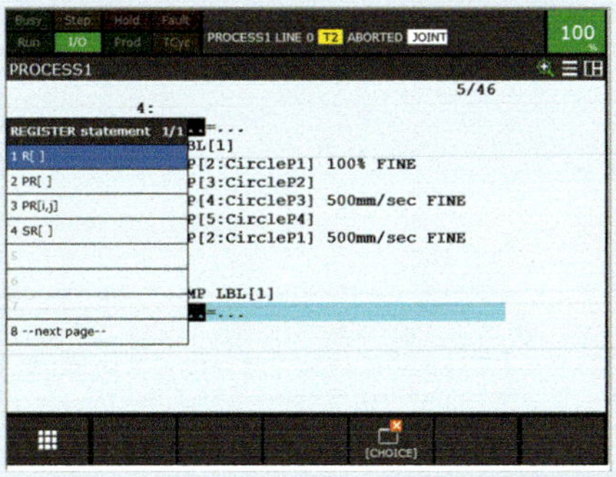

⑫ Type the number of the desired register, set up earlier, for the process count. Then press ENTER. In this case, that number is 1. The second set of dots is highlighted, and the `Register Statement` menu appears.

⑬ Highlight `Constant` and press ENTER.

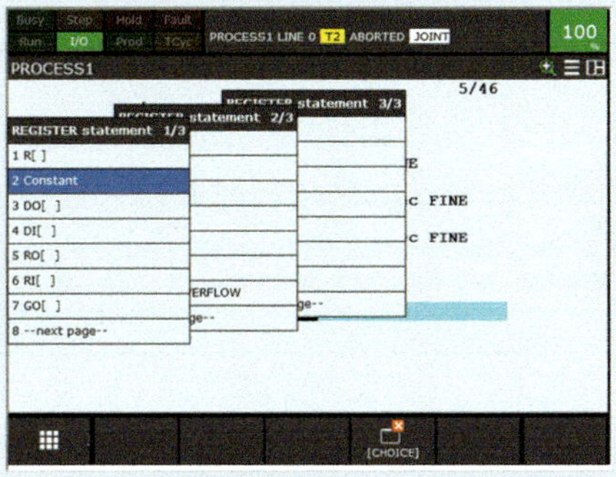

⑭ Type the number 0 and press ENTER.

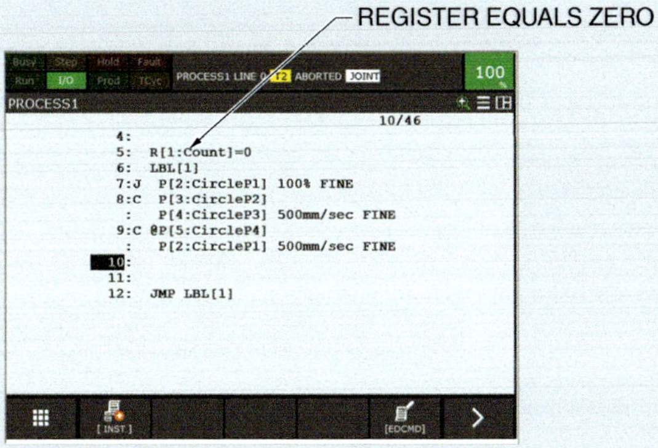

Set the Process Count to Increment by 1

⑮ Highlight a blank line directly after the last line for Process 1. In this example, this is line 10.

⑯ Press F1 `[INST]` to show the `Instruction` menu.

⑰ Highlight `Registers`, and then press ENTER to show the `Register Statement` menu.

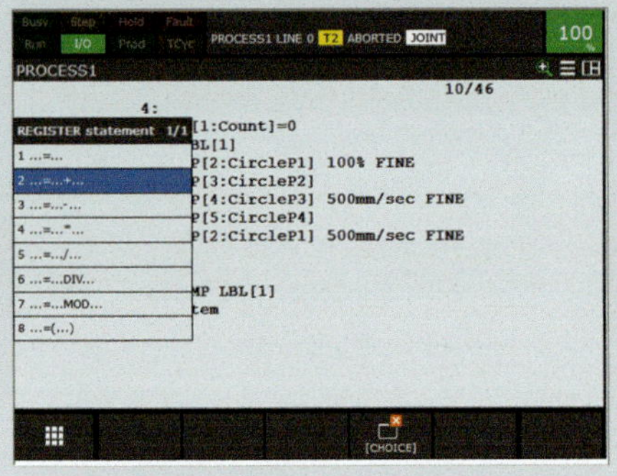

⑱ Highlight ...=...+... and press ENTER. The text ...=...+... appears in the line. The first set of dots is highlighted, and the `Register Statement` menu appears.

⑲ Highlight `R[ ]` and press ENTER. The space between the brackets is highlighted.

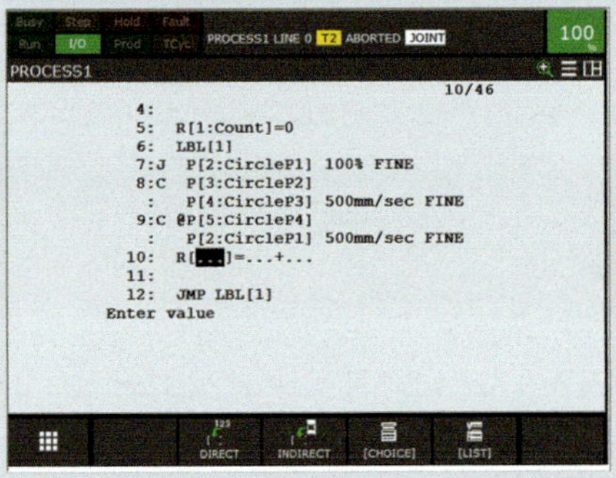

⑳ Type the number of the process count register (1) and then press ENTER. The second set of dots is highlighted, and the `Register Statement` menu appears.

㉑ Highlight `R[ ]` and press ENTER. The space between the brackets is highlighted.

㉒ Type the number of the process count register (1) and then press ENTER. The third set of dots is highlighted, and the `Register Statement` menu appears.

㉓ Highlight `Constant` and press ENTER.

㉔ Type the number 1 and press ENTER.

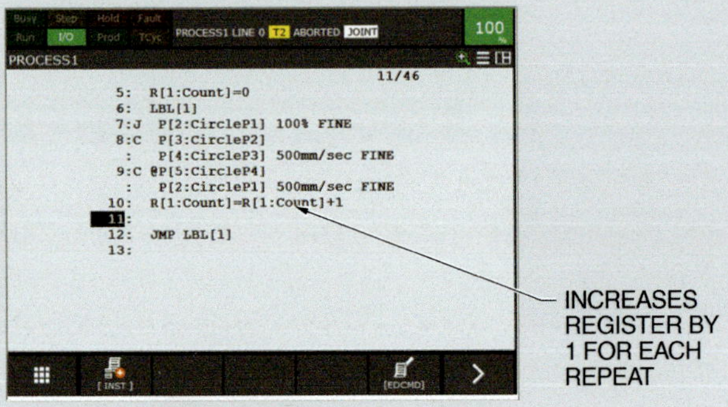

Create an IF/SELECT Statement

(25) Highlight a blank line directly under the register formula line.

(26) Press F1 `[INST]` to show the `Instruction` menu.

(27) Highlight `IF/SELECT` and then press ENTER to show the `IF Statement` menu.

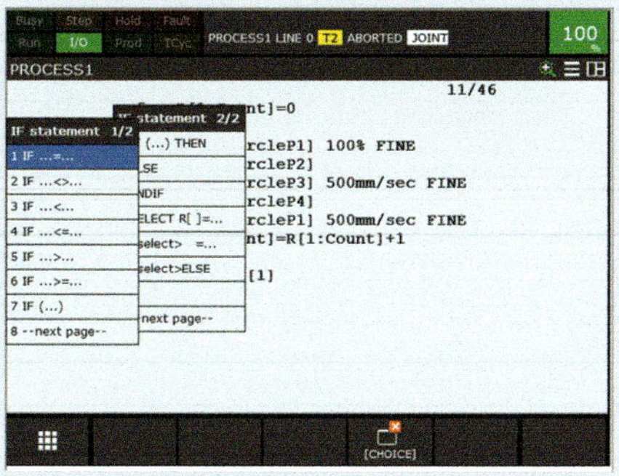

(28) Highlight `IF ...=...` and press ENTER. The text `IF ...=... ...` appears in the line. The first set of dots is highlighted, and the `IF Statement` menu appears.

(29) Highlight `R[ ]` and press ENTER. The space between the brackets is highlighted.

(30) Type the number of the process count register (1) and then press ENTER. The second set of dots is highlighted, and the `IF Statement` menu appears.

(31) Highlight `Constant` and press ENTER.

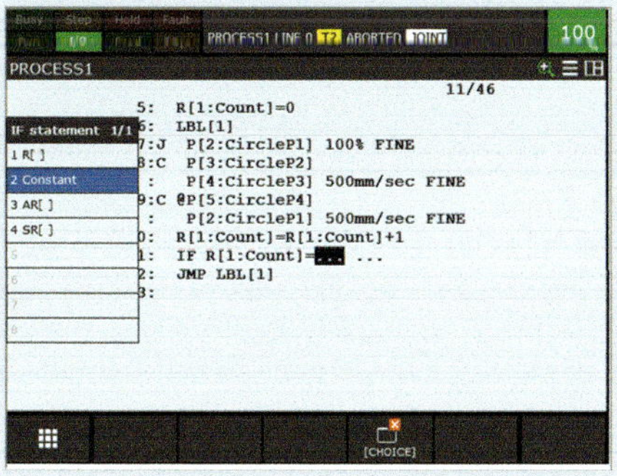

(32) Type the number of times to repeat the code (4) and press ENTER. The third set of dots is highlighted, and the `IF Statement` menu appears.

(33) Highlight `JMP LBL[ ]` and press ENTER. The space between the brackets is highlighted.

(34) Type the number of a new label (2) and press ENTER.

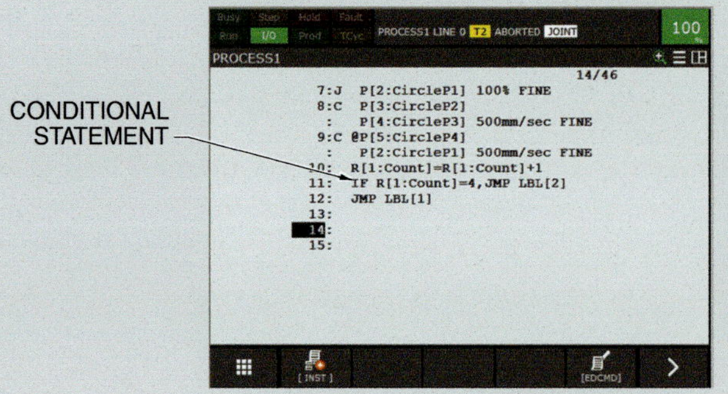

Add a New Label

(35) Highlight a blank line under the `JMP LBL[1]` instruction.

(36) Press F1 `[INST]` to show the `Instruction` menu.

(37) Highlight `JMP LBL/LBL` and press ENTER.

(38) Highlight `LBL[ ]` and press ENTER.

(39) Type the number of the new label (2) and press ENTER.

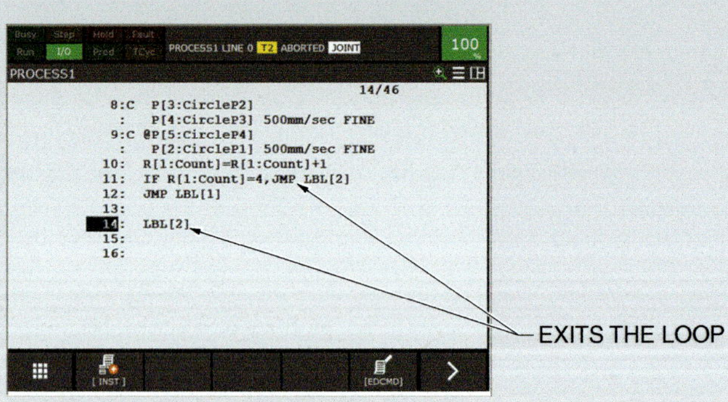

SECTION 7.2 UTILIZING POSITION REGISTERS

Positions are program-specific. This means that the position created in one program cannot be used in another program simply by putting the same position ID in the movement line.

A *position register (PR)* is a position definition saved in memory that can be accessed by any program on a robot. A position consists of a set of numbers—X, Y, and Z values for the three axes and W, P, and R values for the three orientations. Unlike a general-purpose register, which stores one number, a PR stores these six numbers. **See the *Setting Up a Position Register* Procedure.**

PRs are not program-specific. PR numbers are global, which means they can be recalled in any program within the robot. This is useful because sometimes a position must be exactly the same in multiple programs, such as when interfacing with fixed parts in the robot cell. Also, sometimes it is convenient to have some common positions predefined, such as the Home position.

Another advantage of using position registers is that they are not specific to any tool frame or user frame. This means a PR can use the origin, X direction, Y direction, and Z direction of any frame. **See Figure 7-4.** For example, if User Frame 2 is initialized in a sequence of code before a PR is called, the PR uses User Frame 2's origin for positioning the robot. If User Frame 4 is initialized instead, the PR uses the origin for User Frame 4. This feature can be used to efficiently code certain tasks that involve relative locations, such as picking and placing.

Position Registers

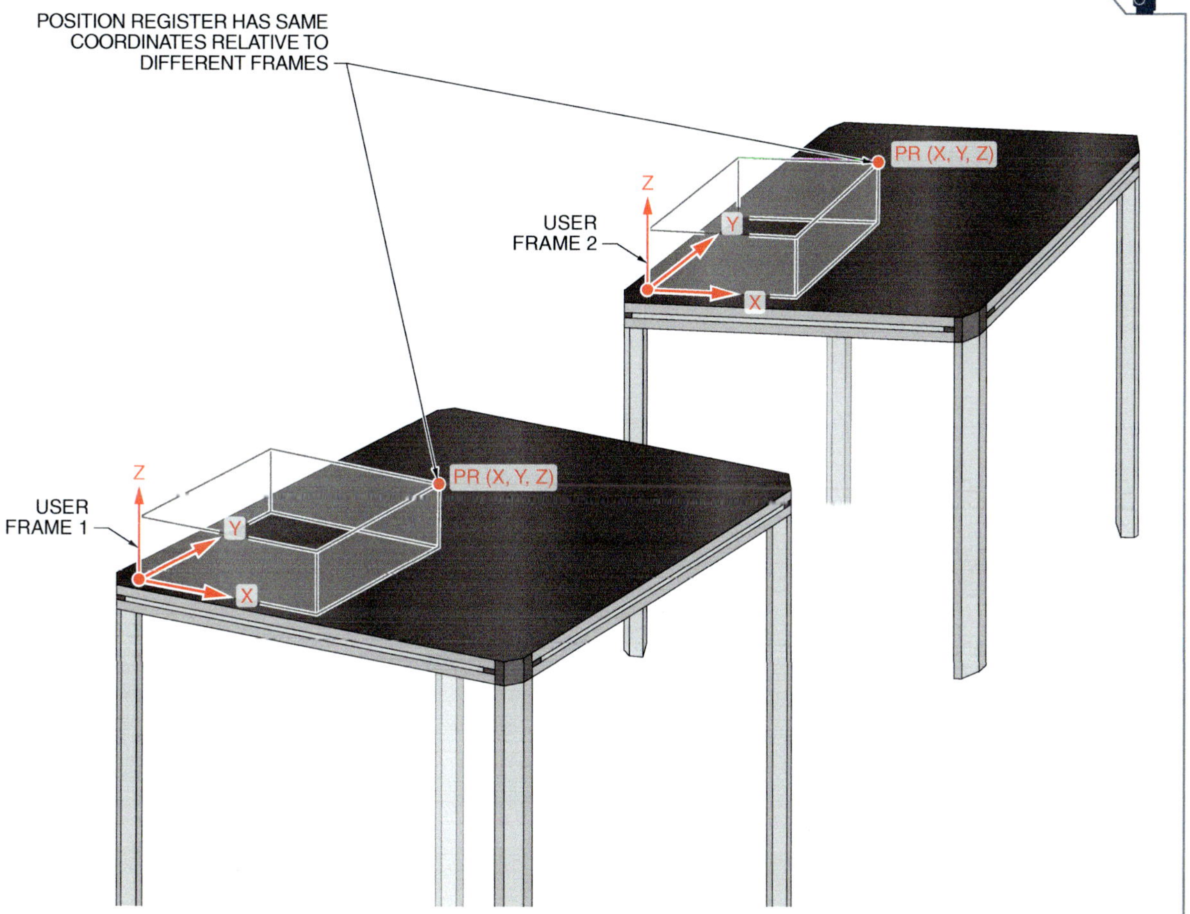

Figure 7-4. Position registers can provide a position relative to any tool or user frame's origin and axes.

PROCEDURE

Setting Up a Position Register

Name the Position Register

① Press DATA.

② Press F1 `[Type]`.

③ Highlight `Position Registers` and press ENTER to show the `DATA Position Reg` screen.

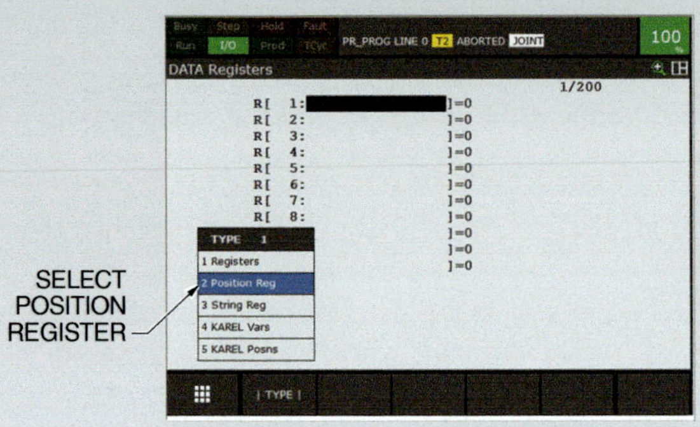

④ Highlight the desired register and press ENTER to add a comment identifying the register.

⑤ Highlight `Options/Keybd`, press F5 `Keyboard`, and type a comment.

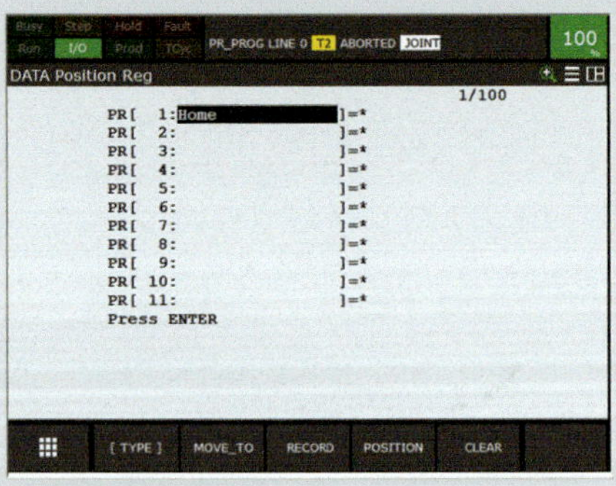

Record the Position

⑥ Jog the robot to the desired position.

⑦ Press and hold SHIFT and then press COORD to open the `Jog` menu.

⑧ Select the desired tool frame and user frame. (For this example, the tool frame is 1 and the Home position is, as always, in User Frame 0 [World frame].)

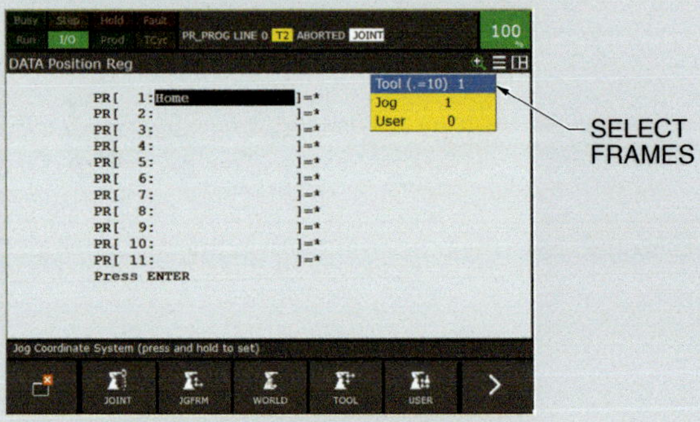

⑨ Press and hold SHIFT and then press F3 `Record`. In the `DATA Position Reg` screen, the * changes to an `R` to indicate that the position is recorded.

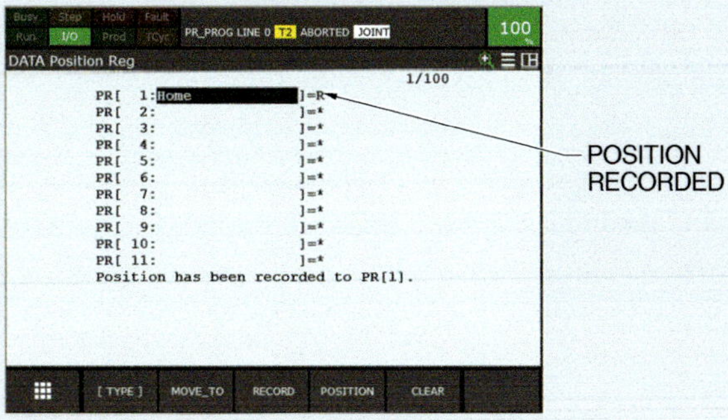

Change a Position to a Position Register

⑩ Open the desired program by pressing SELECT, highlight the program name, then press ENTER.

⑪ Highlight the number of a position.

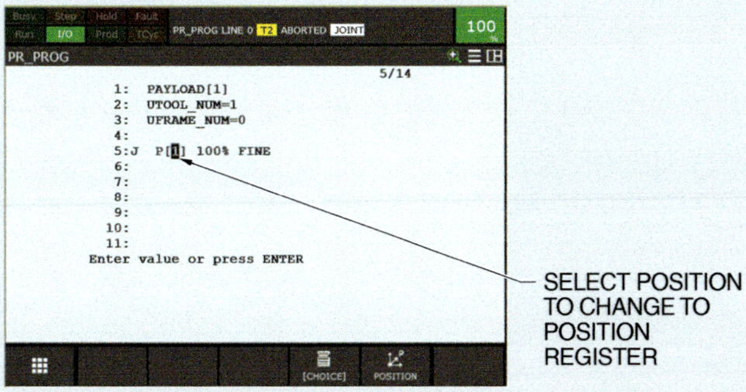

⑫ Press F4 `[Choice]` to show the `Motion Modify` menu.

⑬ Highlight `PR[ ]` and press ENTER. The space between the brackets is highlighted.

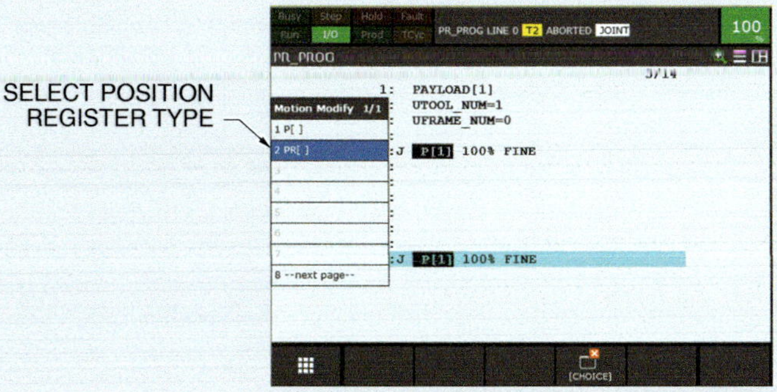

⑭ Type the number of the desired position register and press ENTER.

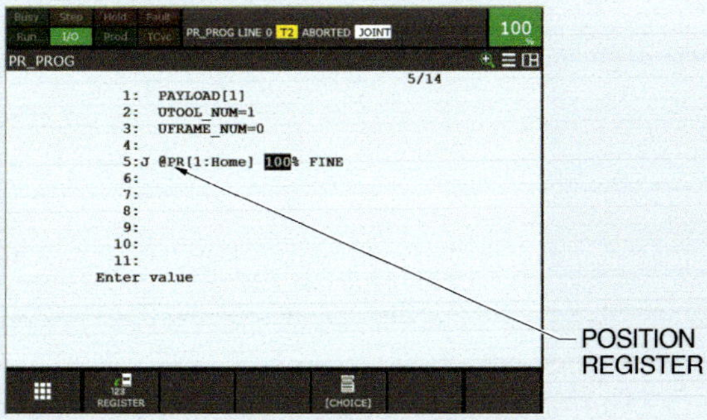

Teach Pendant Software Interfaces

Like all software developers, robot manufacturers continually add features and improve the usability of the robot operating and programming software. For FANUC robots, the update from Version 8 to Version 9 included changes in the teach pendant interface. In Version 8, text menus were the norm, and in Version 9, icons were introduced as the default interface. Many students appear to find the icon-based interface more user-friendly, which helps them to find the instruction quicker and reduces the time required to create a robot program. However, if the menu interface is preferred, the Version 9 system allows the programmer to change between the two.

ICON INTERFACE

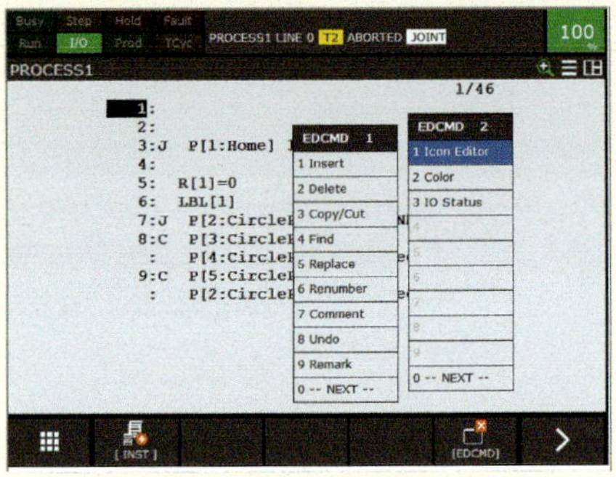

MENU INTERFACE

Changing from the Icon Interface to the Menu Interface

1. Press F5 `[EDCMD]`.
2. Click on the right side arrow to go to the next screen.
3. Click on `EXIT ICON`.

Changing from the Menu Interface to the Icon Interface

1. Press F5 `[EDCMD]`.
2. In the `EDCMD 2` menu, select `Icon Editor`.

Review 7

Name ______________________________ Date ______________

SECTION 7.1 UTILIZING REGISTERS

_______________ **1.** ___ can be used to repeat a section of code, but they have no way of limiting the number of repeats, so they will repeat the section forever.

A. Registers
B. Loopers
C. Jump labels
D. Positions

_______________ **2.** After a register is named, it is typical to set its ___ value (usually zero) at the beginning of a program.

A. initial
B. incremental
C. final
D. target

_______________ **3.** A(n) ___ is a place within the software to store a number.

_______________ **4.** A command line (___ statement) can end the looping when the desired number is reached.

5. What are three ways command lines interact with registers?

6. What is a common application of registers?

SECTION 7.2 UTILIZING POSITION REGISTERS

_______________ **7.** ___ are not program-specific.

A. Labels
B. Positions
C. Position registers
D. Command lines

_______________ **8.** A position consists of a set of ___ numbers.

_______________ **9.** A position register is not specific to any ___ frame or user frame.

_______________ **10.** A position register can use the ___, X direction, Y direction, and Z direction of any frame.

A. position ID
B. frame number
C. name
D. origin

11. Why are global positions useful?

7
Lab Exercise

Lab Objective

Insert required instruction lines to control the repetition of program sections.

Equipment Required

- Robot with end effector, pointer, and associated tool frame
- Modified "SquareTrace" Program from Lab Exercise 6
- "Circle" macro from Lab Exercise 5
- Two square objects, placed in the robot cell, and their associated user frames

Lab Procedure

1. Insert the required lines so that the first square is traced four times before the program continues.
2. Insert the required lines so that the circle and second square are traced two times before the program ends.
3. Insert the required lines so that the entire program is repeated 12 times.
4. Change all instances of the Home position to a position register.

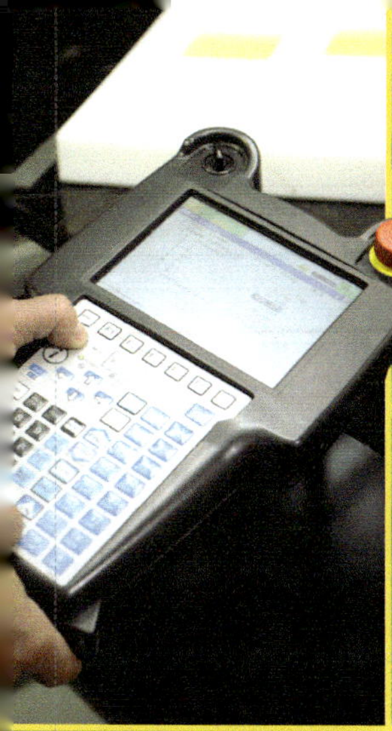

8 Offsetting Position Registers

Previous chapters have used recorded points in order to program a robot's movement. An alternative method is offset programming, which uses the differences in position coordinates to instruct the robot how to move from point to point. If detailed dimensions of the robot cell and the objects the robot interacts with are available, then this programming method can be efficient and precise. A programmer can create that cell in the desktop software's virtual environment. Even planned, future robot cells can be created virtually. This allows the programmer to create much of the program before the robot cell is completed.

Objectives

SECTION 8.1 OFFSET PROGRAMMING

- Explain how offset programming uses position registers.
- Explain the advantage of preprogramming.
- Demonstrate how to code instruction and movement lines for offset programming.
- Demonstrate offset programming using constants.
- Demonstrate offset programming using formulas.
- Explain how to offset multiple coordinates for a single movement.

SECTION 8.2 RESETTING AN OFFSET POSITION REGISTER

- Explain why an offset PR may need to be reset.
- Demonstrate resetting an offset PR using a twin PR.
- Demonstrate resetting an offset PR using registers.

Key Terms

- offset programming
- preprogramming

Learner Resources

atplearningresources.com
Access Code: **762316**

SECTION 8.1 OFFSET PROGRAMMING

Offset programming is the process of programming where position registers (PRs) are modified to become the desired robot positions. Key robot points are created using PRs. Each point is then defined as a change in coordinates from the programmed PR.

For example, a program is created to trace the four corners of a square object, such as a table. For the table's user frame, the origin (0,0,0) is at one corner. With the programming method from previous chapters, the robot would be jogged to each corner, and the points would be recorded. The second point is the next corner at approximately (750,0,0), the third point is approximately (750,500,0), and the fourth point is approximately (0,500,0). It is difficult to get precise coordinates by simply jogging the robot to these positions. **See Figure 8-1.**

Comparing Recorded-Point and Offset Programming

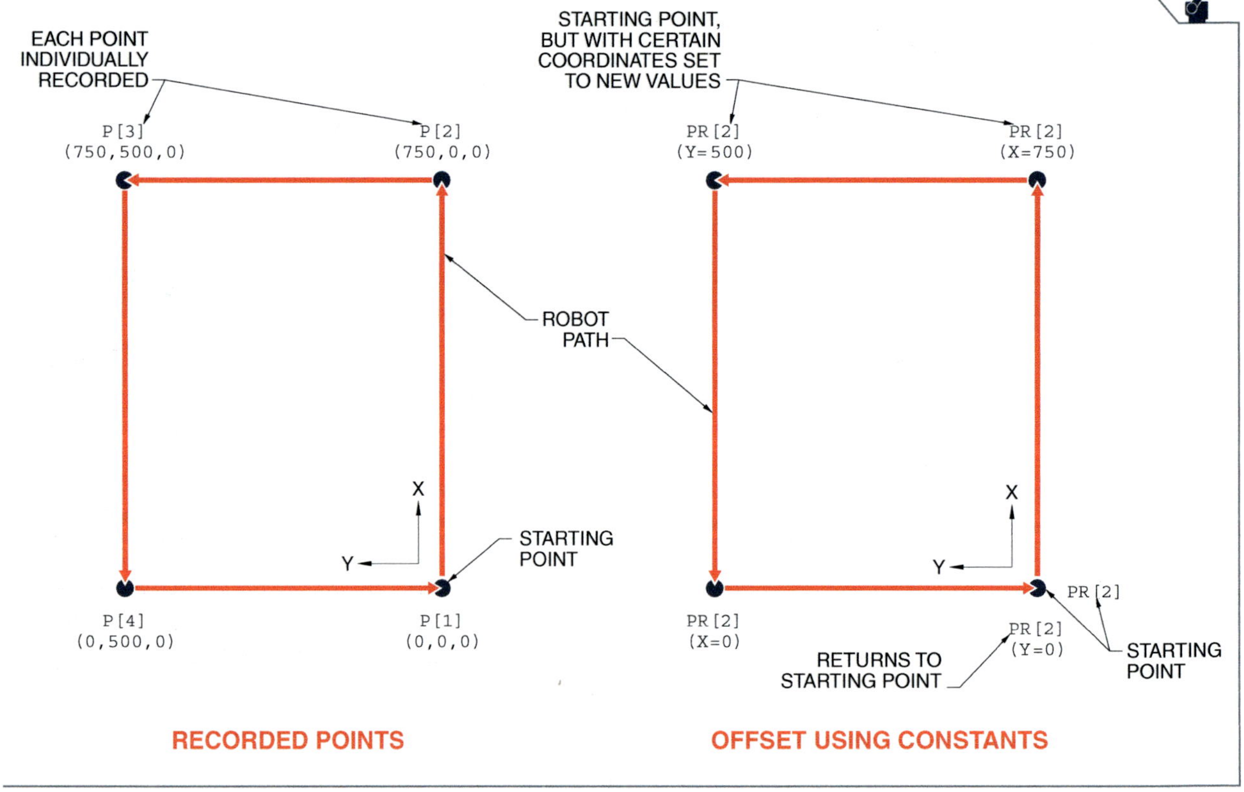

Figure 8-1. The programming method described in earlier chapters requires the robot to move to each point where the position is recorded. Offset programming can be done without a robot and requires only that the values of position and orientation coordinates are changed. This method is typically more precise.

Alternatively, offset programming only requires the programmer to define the values for the position or orientation coordinates that are changing. In the example, from the origin, the second corner would be 750 for the X-axis. The third corner adds 500 for the Y-axis, the fourth returns the X-axis coordinate to 0, and then the Y-axis is set to 0 to bring the robot back to the first corner. Since the coordinates are input into the program, this method provides precise positioning.

PRs are typically used to define the key starting points for offset programming. Their coordinates can be changed easily and individually. Also, since they are not frame-specific, the same PR can potentially be used for interacting with multiple objects. If another square object must be traced and it has its own user frame, then the same PR can be used as a starting point.

Offset Instruction Lines

An offset instruction line is specified in the format `PR[i,j]=`. This specifies the coordinate to be changed and sets up the statement to equal the change. The "i" represents the number of the position register to be used, and the "j" represents the position or orientation coordinate to be changed. This is done by replacing the "j" with a number:

1 = X position coordinate
2 = Y position coordinate
3 = Z position coordinate
4 = W orientation coordinate
5 = P orientation coordinate
6 = R orientation coordinate

This instruction line does not move the robot but just sets a new value in the PR location definition. A subsequent movement line to the PR is used to actually move the robot to the new position.

Dual Check Safety

Dual Check Safety (DCS) is a FANUC software option used to define zones within the robot's operating envelope that require changes in robot operation. DCS can define zones that the robot must not enter, must only enter under certain external conditions (such as no obstructions), or must only enter under certain robot conditions (such as at a reduced speed).

DCS software can be used in many ways. For example, if there is a structural post in the middle of a cell, a zone can be created around the post to prevent the robot from entering. This avoids a potentially damaging crash.

Another example would be to provide safety for personnel who must work near a robot. Using sensors, the robot system can track the locations of people nearby. If a person approaches the robot and enters a certain zone, the robot will run at reduced speed. If the person gets within a closer zone, the robot will stop completely. The DCS software is what enables the use of collaborative robots.

Preprogramming

Creating programs using offsets can only be done if the detailed dimensions of the work area and objects in the robot cell are known. Measurements can be taken in the physical cell if it has already been set up. However, if the cell layout has been pre-engineered and dimensioned drawings are available, then offset programming is easier. Also, the cell and its objects can be created within ROBOGUIDE's virtual robot cell environment. **See Figure 8-2.** The added advantage of doing this is that the robot can then be preprogrammed. *Preprogramming* is the process of creating a robot program entirely within cell simulation software and without access to the physical robot.

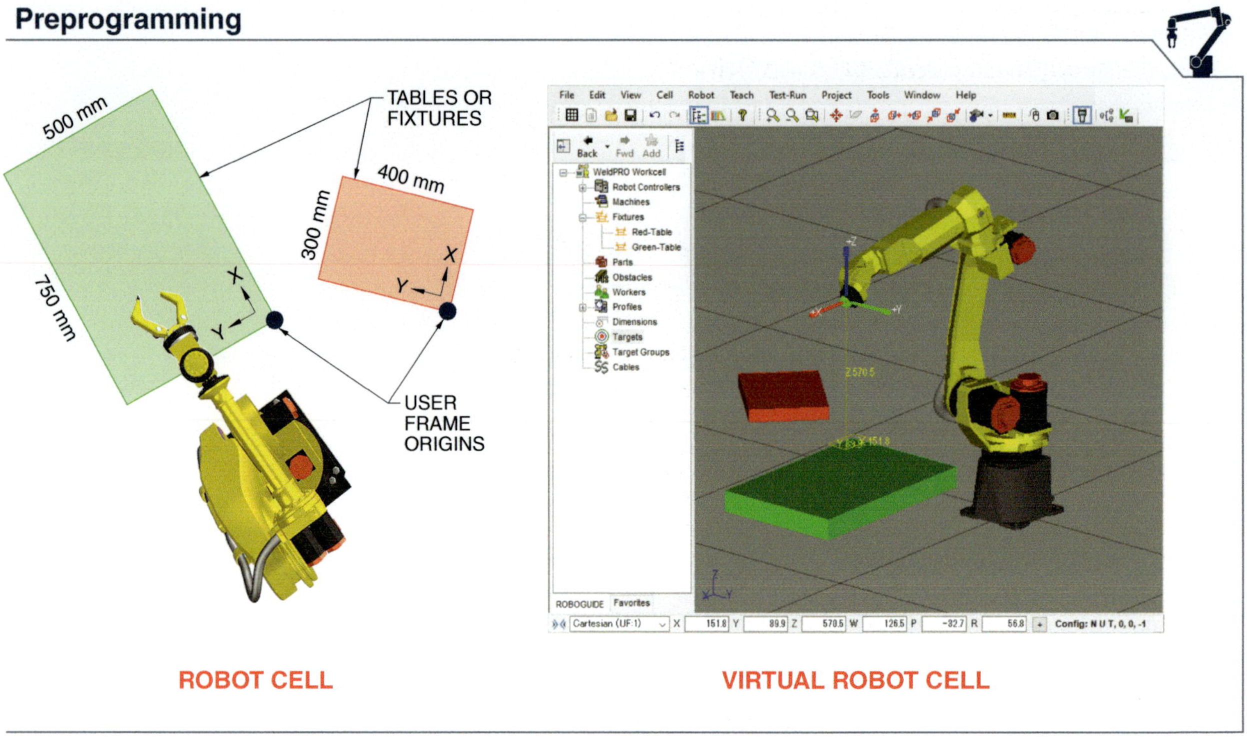

Figure 8-2. Offset programming typically requires detailed dimensions of objects within the robot cell. With this information, the cell can be created as a virtual environment in the ROBOGUIDE software.

Preprogramming allows the bulk of a program to be developed before a robot is installed. Some fine-tuning may be needed later, but most of the programming would be complete. This maximizes productivity because the robot can be put into production soon after being set up.

Offset Programming Using Constants

The simplest method of offset programming is to use an offset instruction line to set one of the position or orientation coordinates to a new value. **See Figure 8-3.** In this example, constant values are used, such as 750 mm in the +X direction. The instruction line for this offset would then be `PR[2,1:Origin]=750`. **See the *Programming with Constant-Value Offsets* Procedure.**

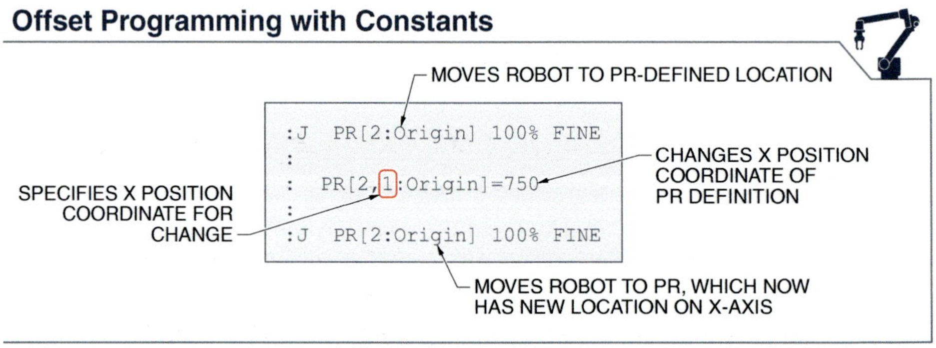

Figure 8-3. Offset programming uses instruction lines that change one of the position and orientation coordinates of a PR and movement lines that move the robot to the modified position.

PROCEDURE

Programming with Constant-Value Offsets

Note: This procedure will trace around the first (green) table. A user frame has already been created for this table.

Move to the User Frame Origin

① Press [MENU], select `SETUP`, and then select `Frames`. Press [ENTER].

② Verify that the User Frame screen is shown. If not, press [F3] `[OTHER]` and select `User Frame`.

③ Highlight the user frame for the first table and press [F2] `DETAIL`.

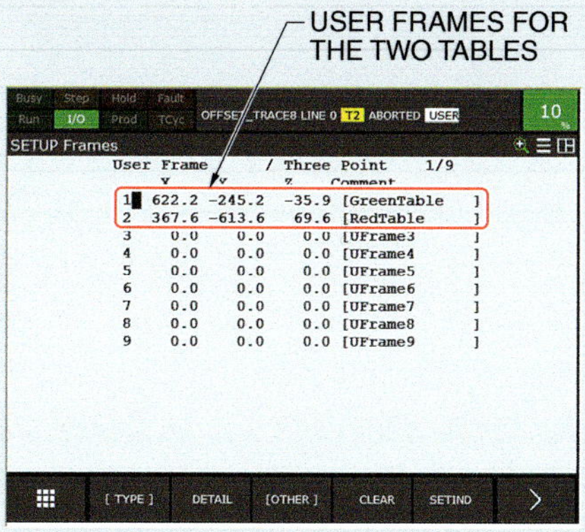

④ Highlight `Orient Origin Point`.

⑤ Press and hold [SHIFT] and then press [F4] `MOVE_TO`. This moves the robot to the origin point for the selected user frame. (Slow to 10% and ensure that the robot has an unobstructed path to the origin.)

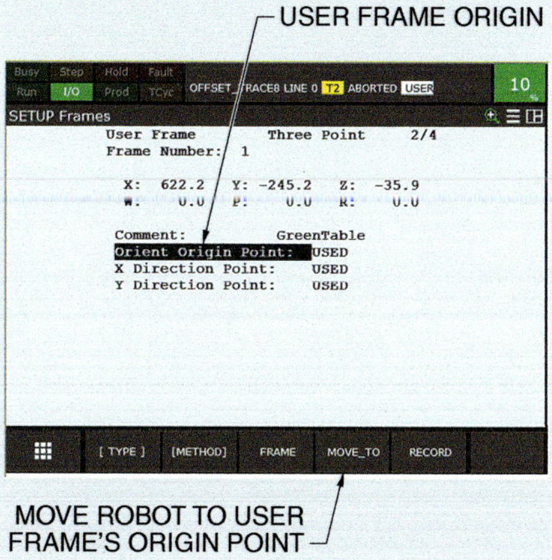

Create the Origin PR

⑥ Open the Jog menu ([SHIFT] and [COORD]) to verify that the correct tool frame and the correct user frame (for the first table) are selected.

⑦ Press [DATA].

⑧ Verify that the DATA Position Reg screen is shown. If not, press [F1] `[TYPE]`, select `Position Reg`, and press [ENTER].

⑨ Highlight the desired PR and press [ENTER] to add a comment, such as "Origin," that identifies the register.

⑩ Press and hold [SHIFT] and then press [F3] `Record`.

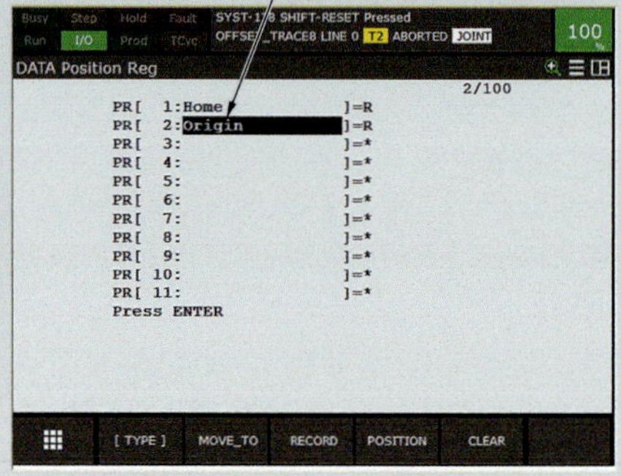

Create the Movement Line to the First Corner

⑪ Add a `UFRAME_NUM=` instruction line to select the user frame for the selected table.

⑫ Highlight the next blank line.

⑬ Press [SHIFT] and then [F1] `Point`.

⑭ Highlight the position number and press [F4] `Choice`.

⑮ Highlight `PR[ ]` and press [ENTER].

⑯ Type the number of the PR (2 in this example) and press [ENTER].

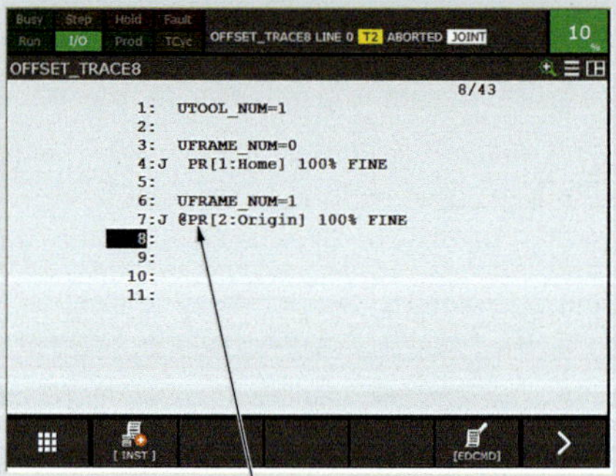

Set the Offset to the Second Corner

⑰ Highlight a blank line under the PR.

⑱ Press [F1] `[INST]`, highlight `Registers`, and press [ENTER].

⑲ Highlight in the menu `...=...` and press [ENTER].

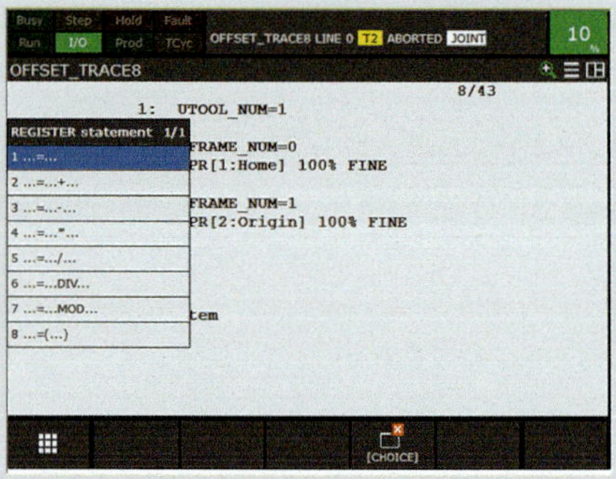

⑳ The first set of dots will be highlighted. Highlight `PR[i,j]` in the menu and press [ENTER].

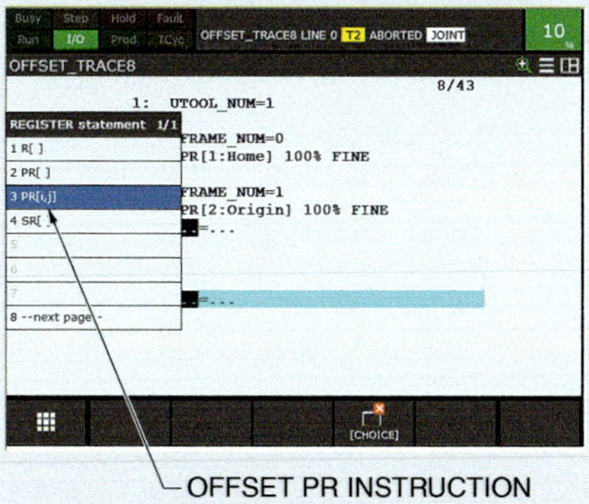

㉑ The `i` is highlighted. Type the number of the PR to be used and press [ENTER]. In this example, the number is 2.

㉒ The `j` is highlighted. Type the number corresponding with the desired position or orientation coordinate offset. In this example, this number is 1 for the X position.

㉓ A menu appears. Highlight `Constant` and press [ENTER].

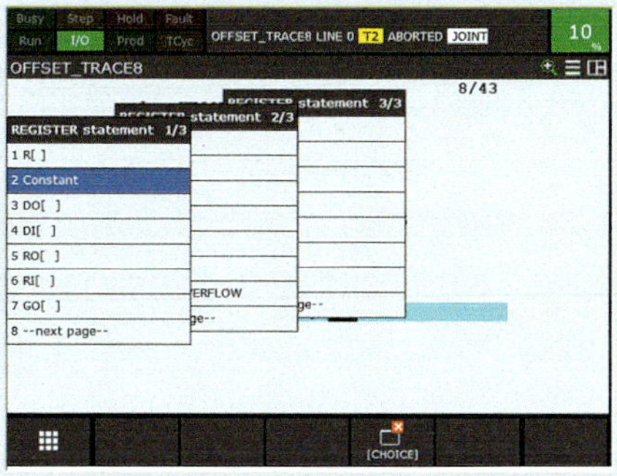

㉔ Type the coordinate value (in mm) and press [ENTER]. In this example, the offset is 750.

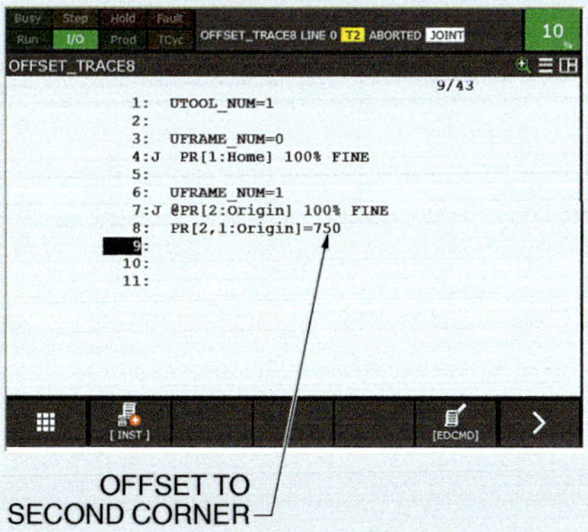

Create the Movement Line to the Second Corner

㉕ Highlight the blank line after the PR offset line.

㉖ Press [SHIFT] and then [F1] `Point`.

㉗ Highlight the position number and press [F4] `Choice`.

(28) Highlight `PR[ ]` and press [ENTER].

(29) Type the number of the PR (2) and press [ENTER].

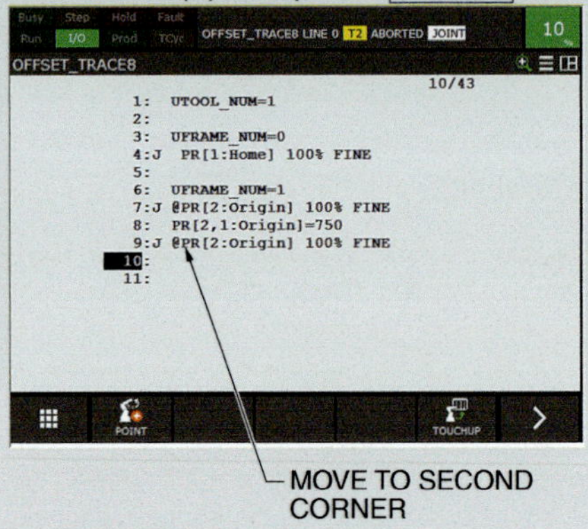

Set the Offset to the Third Corner

(30) Highlight the next blank line.

(31) Repeat steps 18 through 24 to set a new offset but with these values:

- The `i` value is the PR number (2).
- The `j` value is the position/orientation coordinate (2 for the Y position).
- The coordinate value is 500.

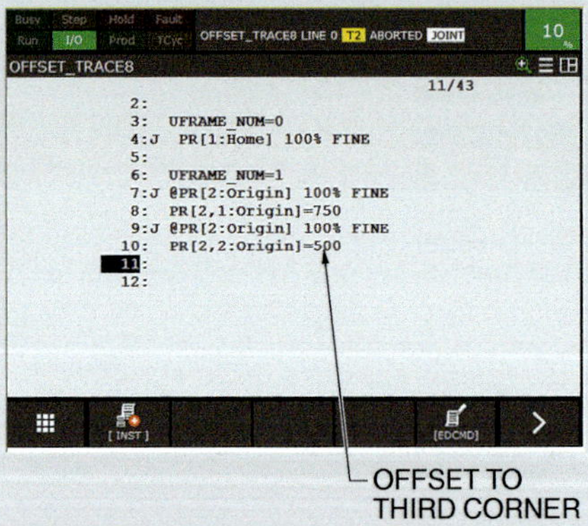

Create the Movement Line to the Third Corner

(32) Highlight the next blank line.

(33) Repeat steps 26 through 29 to add a new movement line. Use the PR number 2.

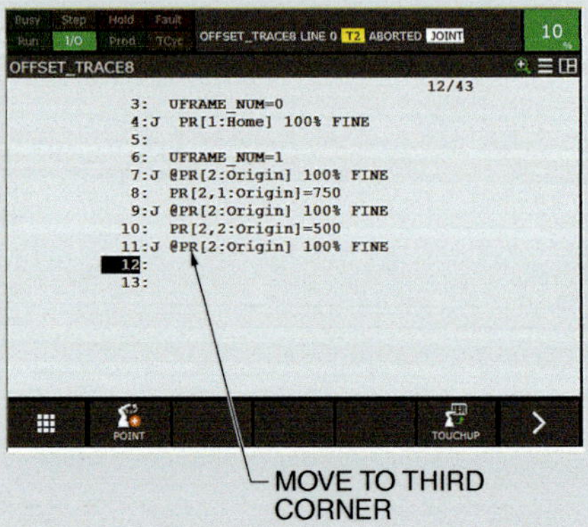

Set the Offset to the Fourth Corner

(34) Highlight the next blank line.

(35) Repeat steps 18 through 24 to set a new offset but with these values:

- The `i` value is the PR number (2).
- The `j` value is the position/orientation coordinate (1 for the X position).
- The coordinate value is 0.

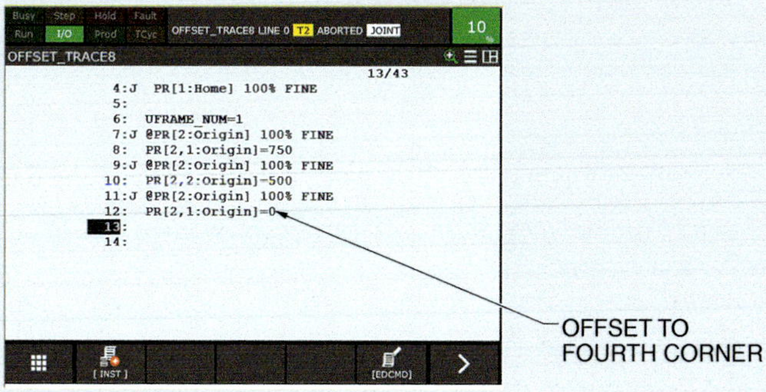

Create the Movement Line to the Fourth Corner

(36) Highlight the next blank line.

(37) Repeat steps 26 through 29 to add a new movement line. Use the PR number 2.

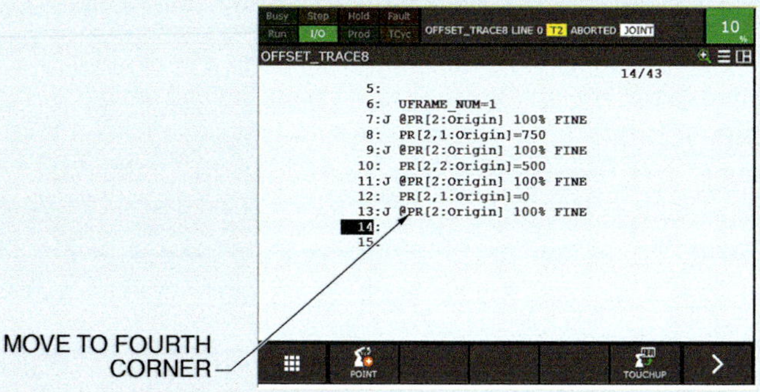

Set the Offset Back to the First Corner

(38) Highlight the next blank line.

(39) Repeat steps 18 through 24 to set a new offset but with these values:

- The `i` value is the PR number (2).
- The `j` value is the position/orientation coordinate (2 for the Y position).
- The coordinate value is 0.

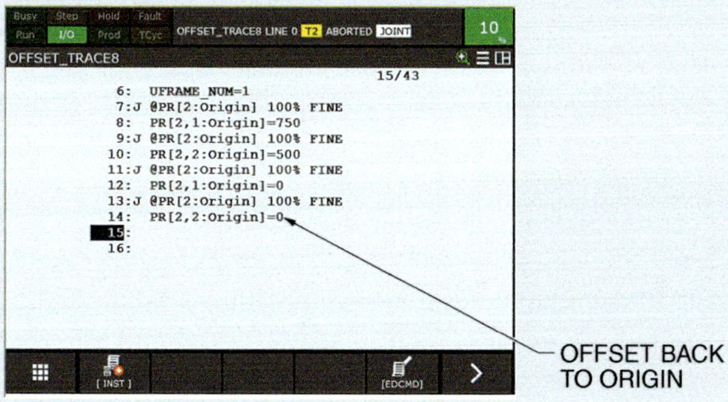

Create the Movement Line Back to the First Corner

(40) Highlight the next blank line.

(41) Repeat steps 26 through 29 to add a new movement line. Use the PR number 2.

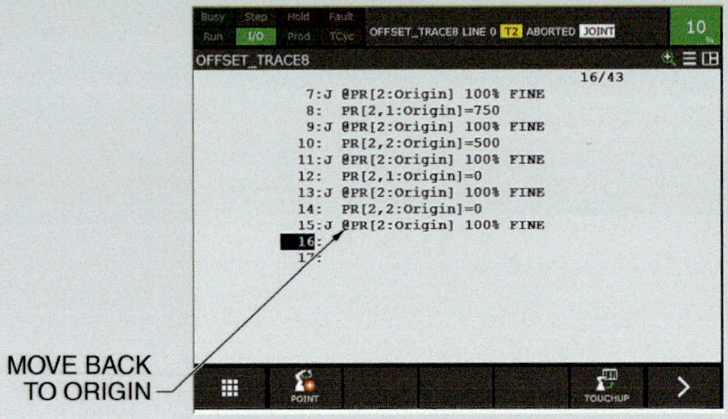

Offset Programming Using Formulas

A more advanced version of offset programming uses math formulas to determine the values of PR coordinates. For some offsets, using a math formula makes for more efficient and more adjustable code. For example, if a robot must interact with equally spaced parts in a grid arrangement, formulas can be used to move the robot in increments from item to item instead of recording each point individually. If the spacing later needs adjustment, changing a formula is likely to be far easier than touching up many points.

A variety of formulas and variables can be used in combination to create complex movements. One of the simplest offsetting formulas is one that changes a PR coordinate in specific increments. **See Figure 8-4.** The instruction line takes the form `PR[i,j]=PR[i,j]+Constant`. For each time this line is executed, the specified PR coordinate increases by a certain amount. **See the *Programming with Formula Offsets* Procedure.**

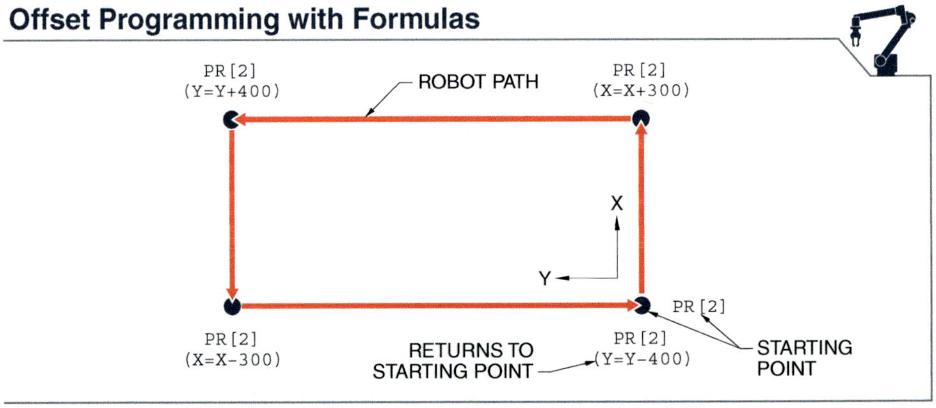

Figure 8-4. Using formulas when offset programming allows the programmer to specify a coordinate change with a mathematical operation.

The same PR number is used for both of the example tables. Each table has its own user frame, and each user frame uses the near-right corner as its origin. Since PRs are not user-frame-specific, the same PR, which is initially defined as the (0,0,0) relative position, can be used. However, it is necessary to select the correct user frame before each table tracing with the instruction line `UFRAME_NUM=`.

PROCEDURE

Programming with Formula Offsets

Note: This procedure will trace around the second (red) table, which is smaller than the first (green) table. This procedure assumes that the origin PR has already been created, along with the necessary user frame.

Create the Movement Line to the First Corner

1. Add a `UFRAME_NUM=` instruction line to select the user frame for the selected table.
2. Highlight the next blank line.
3. Press SHIFT and then F1 `Point`.
4. Highlight the position number and press F4 `Choice`.
5. Highlight `PR[ ]` and press ENTER.
6. Type the number of the offset PR (2 in this example) and press ENTER.

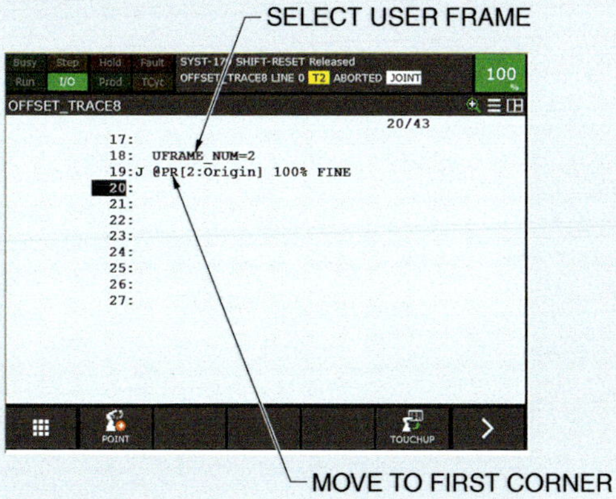

Set the Offset to the Second Corner

7. Highlight the next blank line.
8. Press F1 `[INST]`, highlight `Registers`, and press ENTER.
9. Highlight in the menu `...=...+...` and press ENTER.

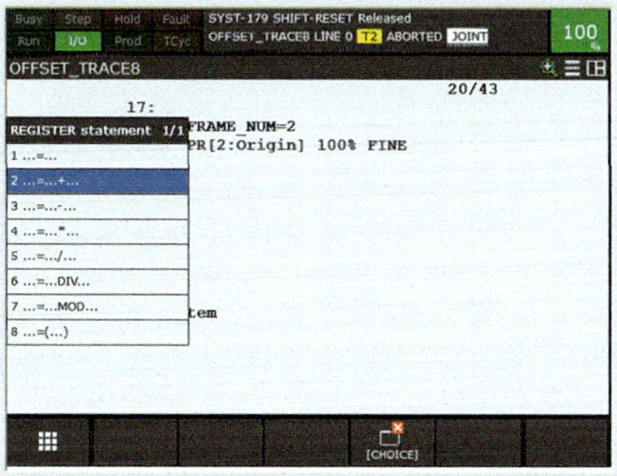

10. The first set of dots will be highlighted. Highlight `PR[i,j]` in the menu and press ENTER.
11. The `i` is highlighted. Type the number of the PR to be used and press ENTER. In this example, the number is 2.

⑫ The `j` is highlighted. Type the number corresponding with the desired position or orientation coordinate offset. In this example, the number is 1 for the X position.

⑬ A menu appears. Select `-- next page --` twice until the third menu page is visible. Highlight `PR[i,j]` and press [ENTER].

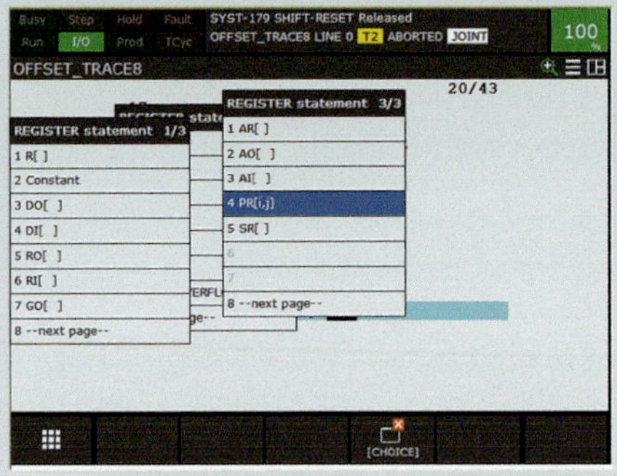

⑭ Type the same values as earlier in the same instruction line:

- The `i` value is the PR number (2).
- The `j` value is the position/orientation coordinate (1 for the X position).

⑮ A menu appears. Highlight `Constant` and press [ENTER].

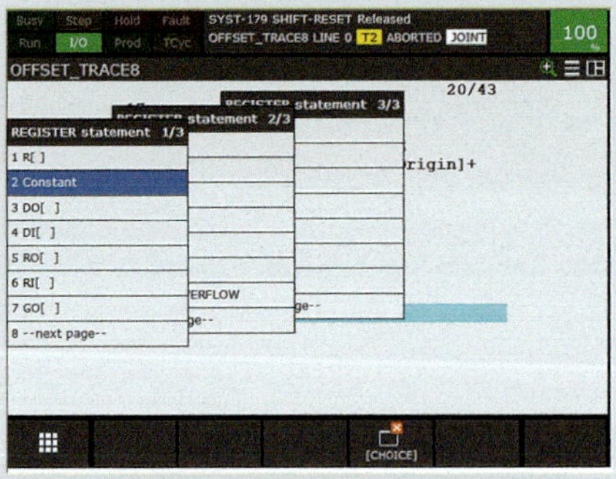

⑯ Type the distance to move (in mm) and press [ENTER]. In this example, the offset is 300.

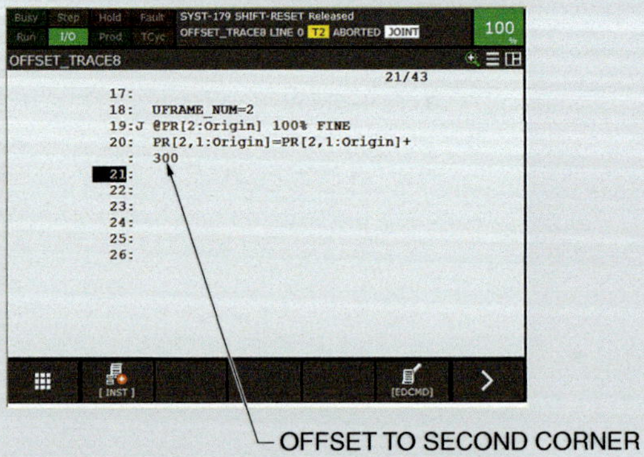

Create the Movement Line to the Second Corner

⑰ Highlight the blank line after the PR offset line.

⑱ Press [SHIFT] and then [F1] `Point`.

⑲ Highlight the position number and press [F4] `Choice`.

⑳ Highlight `PR[ ]` and press [ENTER].

㉑ Type the number of the PR (2) and press [ENTER].

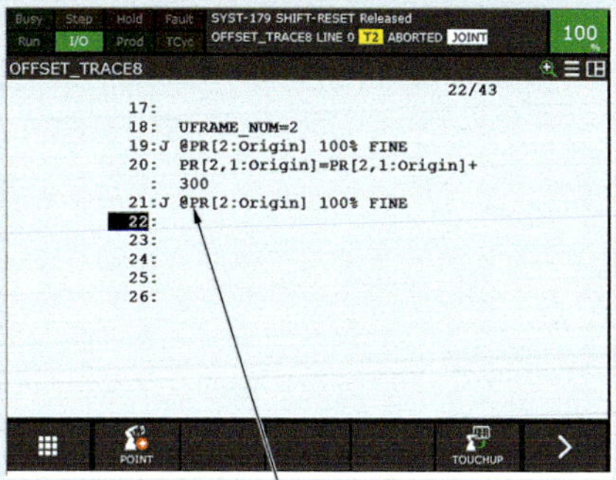

MOVE TO SECOND CORNER

Set the Offset to the Third Corner

㉒ Highlight the next blank line.

㉓ Repeat steps 8 through 16 to set a new offset but with these values:

- The `i` value is the PR number (2).
- The `j` value is the position/orientation coordinate (2 for the Y position).
- The distance to move is 400.

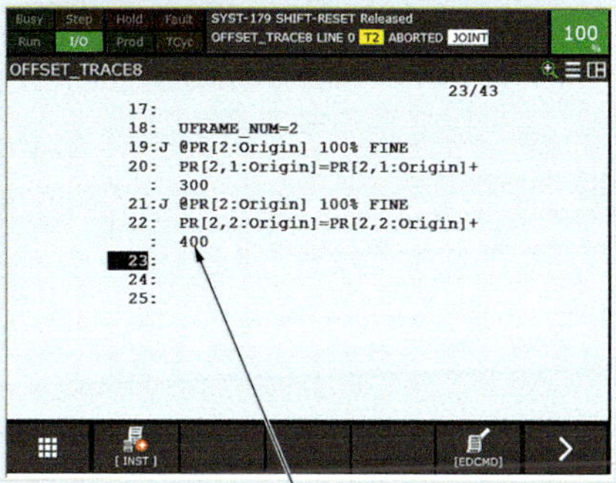

OFFSET TO THIRD CORNER

Create the Movement Line to the Third Corner

㉔ Highlight the next blank line.

㉕ Repeat steps 18 through 21 to add a new movement line. Use the PR number 2.

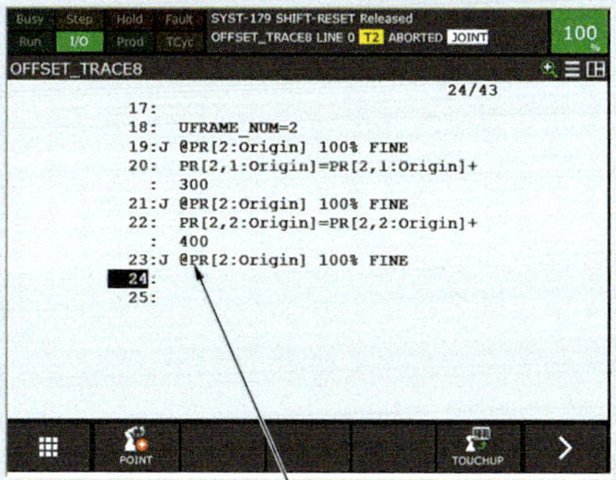

MOVE TO THIRD CORNER

Set the Offset to the Fourth Corner

(26) Highlight the next blank line.

(27) Repeat steps 8 through 16 to set a new offset but with these values:

- Select the `...=...-...` instruction. (Note the subtraction operation.)
- The `i` value is the PR number (2).
- The `j` value is the position/orientation coordinate (1 for the X position).
- The distance to move is 300.

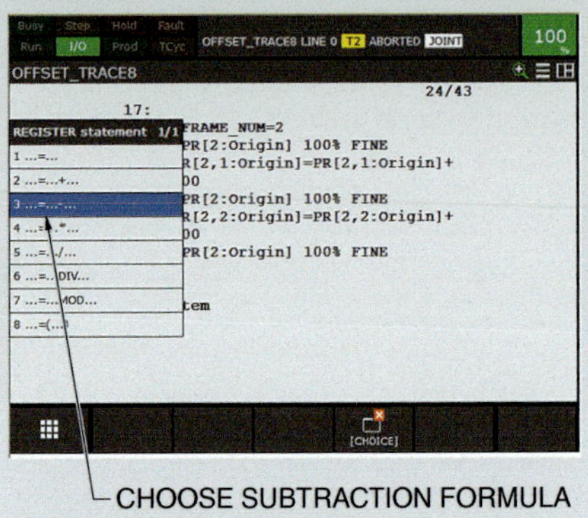

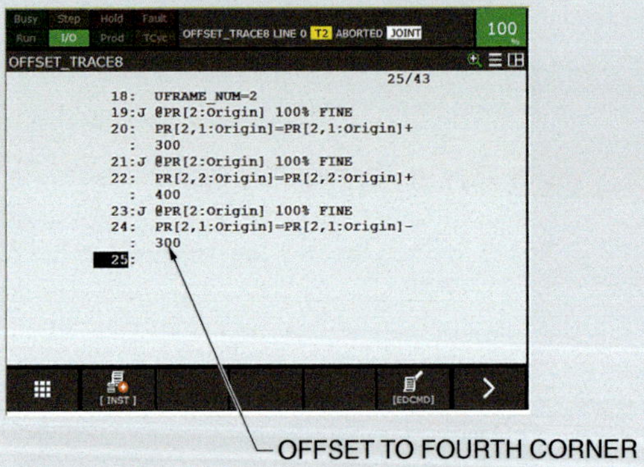

Create the Movement Line to the Fourth Corner

(28) Highlight the next blank line.

(29) Repeat steps 18 through 21 to add a new movement line. Use the PR number 2.

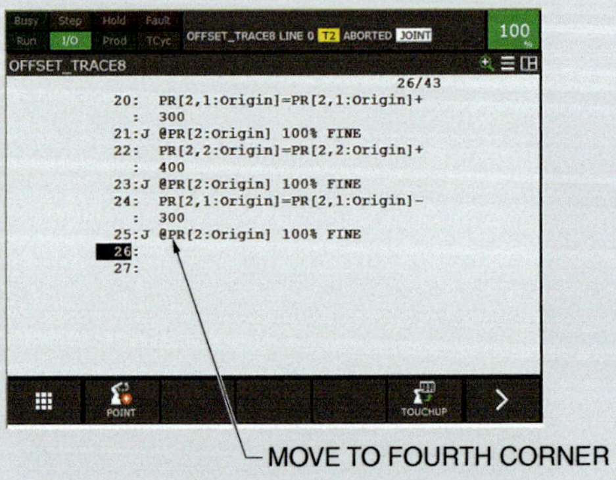

Set the Offset Back to the First Corner

(30) Highlight the next blank line.

(31) Repeat steps 8 through 16 to set a new offset but with these values:

- Select the `...=...-...` instruction. (Note the subtraction operation.)
- The `i` value is the PR number (2).
- The `j` value is the position/orientation coordinate (2 for the Y position).
- The distance to move is 400.

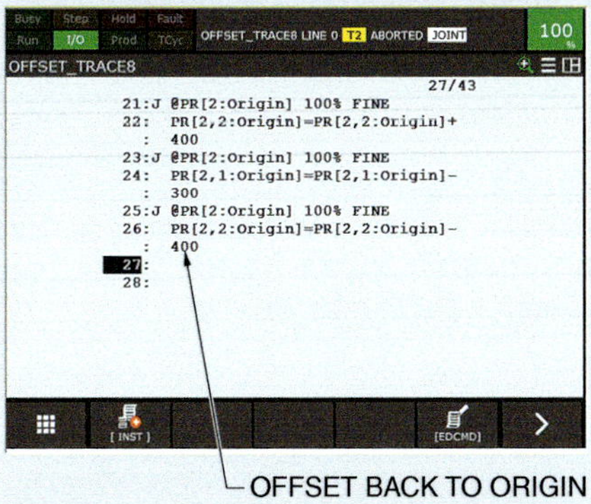

Create the Movement Line Back to the First Corner

(32) Highlight the next blank line.

(33) Repeat steps 18 through 21 to add a new movement line. Use the PR number 2.

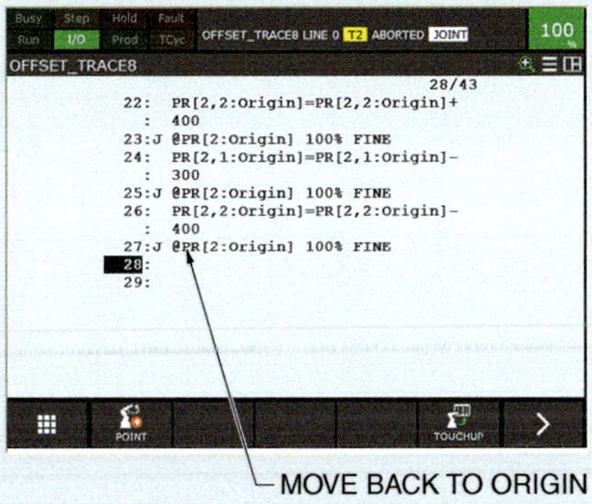

Stacking Offsets

In the table-tracing examples, only one axis coordinate was changed for each move. In other programs, it may be necessary to change multiple coordinates when moving from one position to another. Each instruction line can change only one coordinate of the PR. However, multiple offset instruction lines can be stacked together for a single movement line. **See Figure 8-5.** Up to all six of the PR coordinates (three positions and three orientations) can be changed at a time.

Stacking Offsets

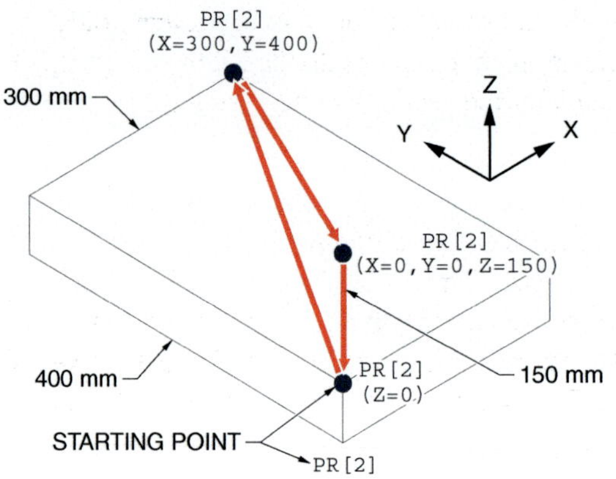


```
:J  PR[2:Origin] 100% FINE
:   PR[2,1:Origin]=300
:   PR[2,2:Origin]=400
:L  PR[2:Origin] 100% FINE
:   PR[2,1:Origin]=0
:   PR[2,2:Origin]=0
:   PR[2,3:Origin]=150
:L  PR[2:Origin] 100% FINE
:   PR[2,3:Origin]=0
:L  PR[2:Origin] 100% FINE
```

Figure 8-5. If a movement requires a change in more than one PR coordinate, multiple offset instruction lines can be put before the movement line.

SECTION 8.2 RESETTING AN OFFSET POSITION REGISTER

Since, in both examples, the robot ended the table tracing programs where it started, the PR was reset to the original value. This is not the case for all programs. When the offset instruction is used on a PR, the PR is changed permanently and alters the position of the robot. If the program does not return the PR to the starting point and the program is executed a second time, the PR will likely not be in the proper position.

To verify that the PR is back at the origin of the user frame, the programmer can view the PR information in the DATA screen. **See the *Verifying the Offset PR Position* Procedure.** In this example, the PR position coordinates are expected to be 0,0,0.

PROCEDURE

Verifying the Offset PR Position

1. Press DATA.
2. Verify that the DATA Position Reg screen is shown. If not, press F1 [TYPE], select Position Reg, and press ENTER.
3. Highlight the PR used as the offset PR. In this example, it is PR[2].
4. Press F4 POSITION. The position (X, Y, and Z) and orientation (W, P, and R) information is shown at the top of the screen.

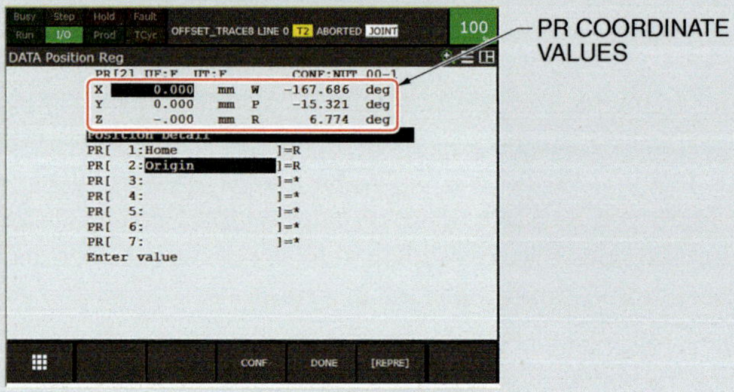

Since the program is likely to be executed multiple times, it is typically necessary to reset the PR to its original position if it is not returned there by the program. There are two common methods for doing this: copying the coordinates from a twin PR that holds the original position or copying the original position coordinates into the PR from a group of registers. These sections of code are placed after the section of code using the offset PR.

Resetting the Offset PR Using a Twin PR

One way to reset the PR is by creating two PRs that are exactly the same (twin PRs). Both would describe the position at the origin of the user frame. Over the course of the program, the first PR is changed with offsets while the second (twin) PR remains the same. At the end of the program, an instruction line is used to set the position of the offset PR to the value of the twin PR. **See the *Resetting the Offset PR with a Twin PR* Procedure.** This effectively resets the offset PR to its original position.

PROCEDURE

Resetting the Offset PR with a Twin PR

1. Create a second PR in the same manner as the first. It must be also defined as the origin of the first table's user frame. Name the second PR similarly to the first but with an indication that it is a duplicate, such as "Origin-Twin." Refer to steps 1 through 10 in the *Programming with Constant-Value Offsets* procedure.

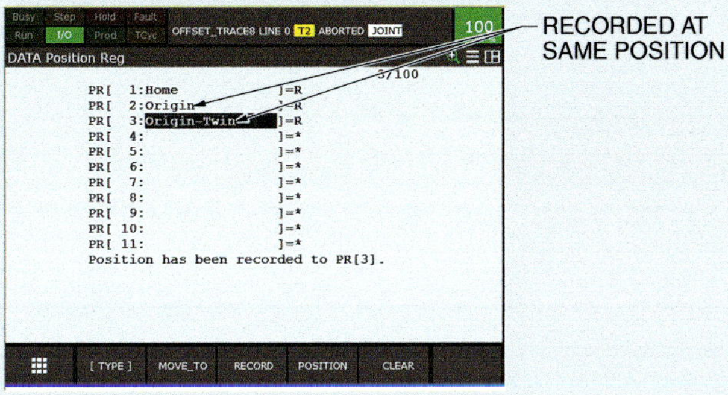

2. In the program, after the code tracing the table, highlight the next blank line.
3. Press F1 `[INST]`, highlight `Registers`, and press ENTER.
4. Highlight `...=...` and press ENTER.
5. Highlight `PR[ ]` and press ENTER.
6. Type the number of the PR used for offsets (2 in this example) and press ENTER.
7. Highlight `PR[ ]` and press ENTER.
8. Type the number of the PR (such as 3) used as the twin and press ENTER.

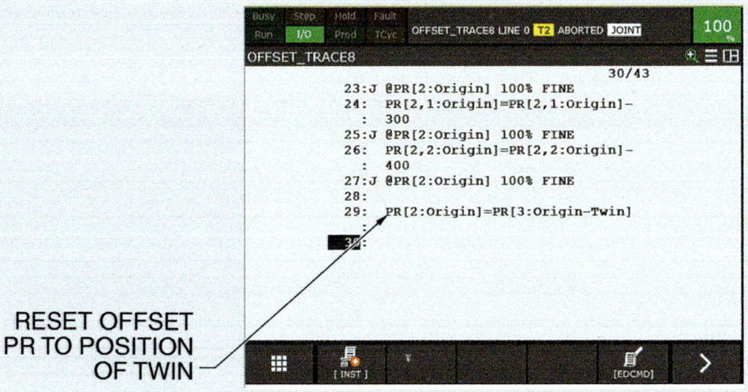

Resetting the Offset PR with Registers

Another method for resetting the PR position is by using registers. Since there are six position coordinates (three position and three orientation), six registers must be available to reset a PR. In this example, the values are saved into registers 10 through 15. Then in the program, an instruction line is used to copy a register value into the PR for each of the six coordinates. **See the *Resetting the Offset PR with Registers* Procedure.**

For the position coordinates, the desired values will likely be known, such as (0,0,0) for an origin PR. For the orientation coordinates, if the necessary values are not already known, the programmer might use the angles for the PR shown in Position Reg screen.

PROCEDURE

Resetting the Offset PR with Registers

Copy the PR Coordinates into Registers

1. Press DATA.
2. Press F1 `[TYPE]`, highlight `Registers`, and press ENTER.
3. Enter the desired values into the selected registers. Then press ENTER.

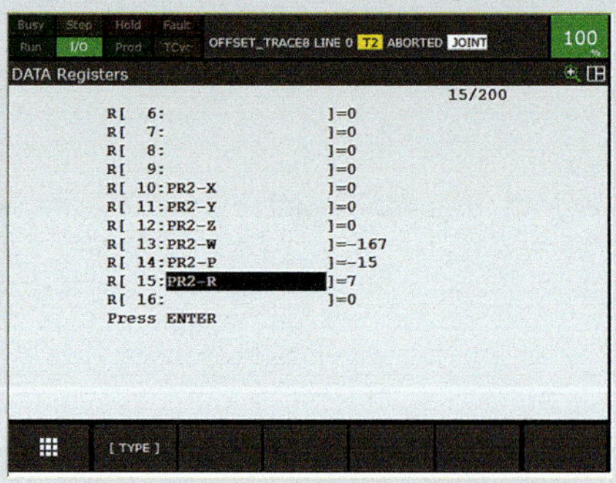

4. Open the program in which the PR will be reset.
5. At a line just after the code using the offset PR, insert six sequential blank lines.
6. Highlight the line number of the first blank line.

Set the Value of the First PR Coordinate

7. Press F1 `[INST]`, highlight `Registers`, and press ENTER.
8. Highlight . . . = . . . and press ENTER.
9. The first set of dots will be highlighted. Highlight `PR[i,j]` in the menu and press ENTER.
10. The `i` is highlighted. Type the number of the PR to be used and press ENTER. In this example, the number is 2.
11. The `j` is highlighted. Type the number corresponding with the desired position or orientation offset. In this example, this number is 1 for the X position.
12. A menu appears. Highlight `R[ ]` and press ENTER. The space between the brackets is highlighted.

⑬ Type the number of the register that corresponds to the position or orientation to be changed. In this example, Register 10 corresponds to the X position. If needed, press F5 `[LIST]` for a list of registers and their identifying comments.

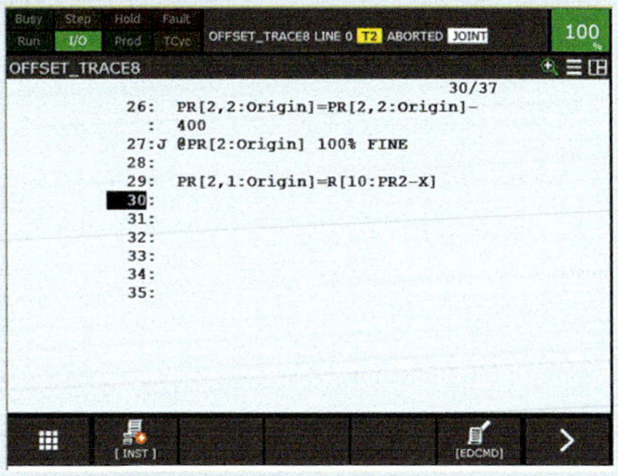

⑭ Repeat steps 7 through 13 for the other positions and orientations.

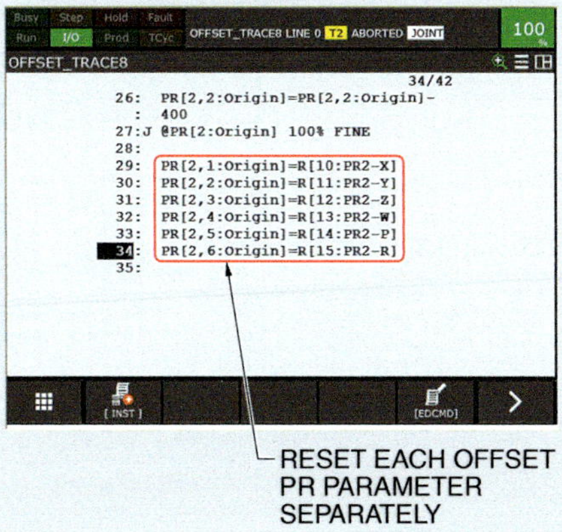

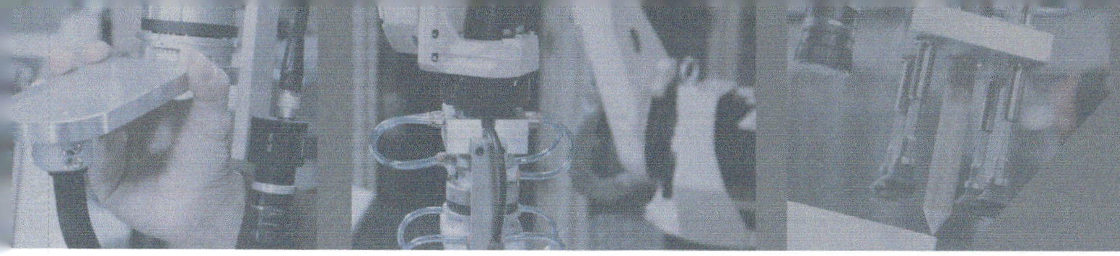

Review 8

Name ______________________________ Date ____________________

SECTION 8.1 OFFSET PROGRAMMING

________ **1.** Since coordinates are input into the program, ___ programming provides precise positioning.

________ **2.** ___ are typically used to define the key starting points for offset programming.

A. Registers
B. Position registers
C. Home positions
D. World origins

________ **3.** Since they are not ___-specific, the same PR can potentially be used for interacting with multiple objects.

________ **4.** The "i" in `PR[i,j]` represents the ___.

A. number of the robot group
B. distance to move
C. axis direction to move
D. number of the position register

________ **5.** The "j" in `PR[i,j]` represents the ___.

A. number of the robot group
B. distance to move
C. number of the position or orientation to change
D. position register

________ **6.** An instruction line does not move the robot. A subsequent ___ to the PR is used to actually move the robot to the new position.

________ **7.** For some offsets, using a math formula makes for more efficient and more ___ code.

A. adjustable
B. compact
C. readable
D. durable

____________________ **8.** It is necessary to select the correct ___ frame before each table tracing.

____________________ **9.** When the offset instruction is used on a PR, the PR is changed ___.

A. permanently
B. until the program restarts
C. until the controller reboots
D. for one cycle

10. Describe the programming method from previous chapters.

11. What must the programmer define when using the offset programming method?

12. What is an advantage to preprogramming?

13. How do detailed dimensions of a robot cell help offset programming?

14. How are multiple coordinates changed for an offset PR?

SECTION 8.2 RESETTING AN OFFSET POSITION REGISTER

____________________ **15.** To verify that the PR is back at the origin of the user frame, the PR information is viewed in the ___ screen.

____________________ **16.** Over the course of a program, the first PR is changed with offsets while the second (___) PR remains the same.

____________________ **17.** To reset a PR, ___ register(s) must be available.

A. 1

B. 3

C. 6

D. 10

18. When is it typically necessary to reset a PR to its original position?

8 Lab Exercise

Lab Objectives

- Demonstrate how position registers are affected by user frame selection.
- Insert required instruction lines to offset a position register in a pattern.

Equipment Required

- Robot with end effector, pointer, and associated tool frame
- Two square objects, placed in the robot cell, and their associated user frames

Lab Procedure

1. Start a new program shell.
2. Create a Home position register (number 1) in World.
3. Create a position register (number 2) at a corner of one of the squares. Ensure that the position is created in the proper user frame.
4. Create the following program. Run the program and observe how the user frame selection affects the position register location.

```
:  UTOOL_NUM=1
:  UFRAME_NUM=0
:J  PR[1:Home] 100% FINE
:
:  UFRAME_NUM=1
:J  PR[2:Origin] 100% FINE
:
:  UFRAME_NUM=2
:J  PR[2:Origin] 100% FINE
```

5. Trace each of the two squares using only the PR and offsets. The program in the previous step can be used as a starting point.

9

Inputs/Outputs and End Effectors

An industrial robot may need to interact with processes both ahead of and after it in order to keep the production line moving smoothly. This typically requires reading input information, sending output signals, and incorporating both into the robot program for coordinating tasks between machines. The part of a robot that either holds an object or involves tooling is the end effector, which is not included when a robot is purchased. An appropriate end effector is chosen for each specific task the robot needs to perform or product the robot needs to hold. There are many different types of end effectors and options for attaching them.

Objectives

SECTION 9.1 UTILIZING INPUTS/OUTPUTS

- Differentiate between inputs and outputs and between digital signals and analog signals.
- Describe the primary use for robot I/O.
- Demonstrate how to initiate and program robot I/O.
- List some common uses for digital I/O.
- Demonstrate how to use digital I/O in a TP program.
- Describe the purpose of group I/O.
- List some possible analog I/O applications.

SECTION 9.2 INTEGRATING END EFFECTORS

- Describe the primary considerations when choosing or designing an end effector.
- Describe the operation of pneumatic and gripper-type end effectors.
- Compare the rigid attachment of end effectors with using tool changers.
- Explain the importance of part-present devices.
- Describe three types of part-present devices and their operation.

Key Terms

- input
- output
- digital signal
- analog signal
- robot I/O
- complementary outputs
- digital I/O
- group I/O
- analog I/O
- stripped-down robot
- integrator
- repeatability
- combination end effector
- rigid attachment
- tool changer
- part-present device
- limit switch
- proximity switch
- pressure switch

Learner Resources
atplearningresources.com
Access Code: **762316**

SECTION 9.1 UTILIZING INPUTS/OUTPUTS

Consider a robot programmed to load and unload parts for a machining center. To keep the machining center producing parts efficiently, the robot must do all of the following:

- Detect the part that is next to be loaded into the machining center.
- Monitor for any alarms from the machining center.
- Determine if the machining center door is open or closed.
- Grasp and move the part.
- Verify that the part is loaded correctly into the machining center's fixture.
- Receive notification when the part is finished.

The robot cannot "see" the location of the parts, when a part is completed, if a part is loaded properly, or if the machining center's door is open or closed. To overcome these issues, a FANUC robot is equipped with four input (I) and output (O) options: robot I/O, digital I/O, group I/O, and analog I/O. These inputs and outputs allow the robot to receive status information and make changes to cell equipment.

An *input* is an information signal that is supplied to a controller from other equipment. Input signals typically come from sensors reacting to some aspect of the environment, such as a door switch closing. An *output* is a control signal supplied from a controller to other equipment. Output signals are commonly used to open or close valves or to control the speed of rotating equipment.

Inputs and outputs can be either digital or analog. **See Figure 9-1.** A *digital signal* is an electrical signal that indicates only ON or OFF (or HIGH or LOW) states. A common type of digital signal defines an OFF as a near-zero voltage and an ON as a voltage within a small range. An *analog signal* is an electrical signal that varies in voltage or current in accordance with some variable parameter. The range and the resolution (the size of incremental steps) of analog signals can vary for different systems or equipment. For example, one analog signal may cover a range of 0 V to 5.0 V, with a resolution of 0.1 V. For safety, all types of signal voltages are generally less than 50 V.

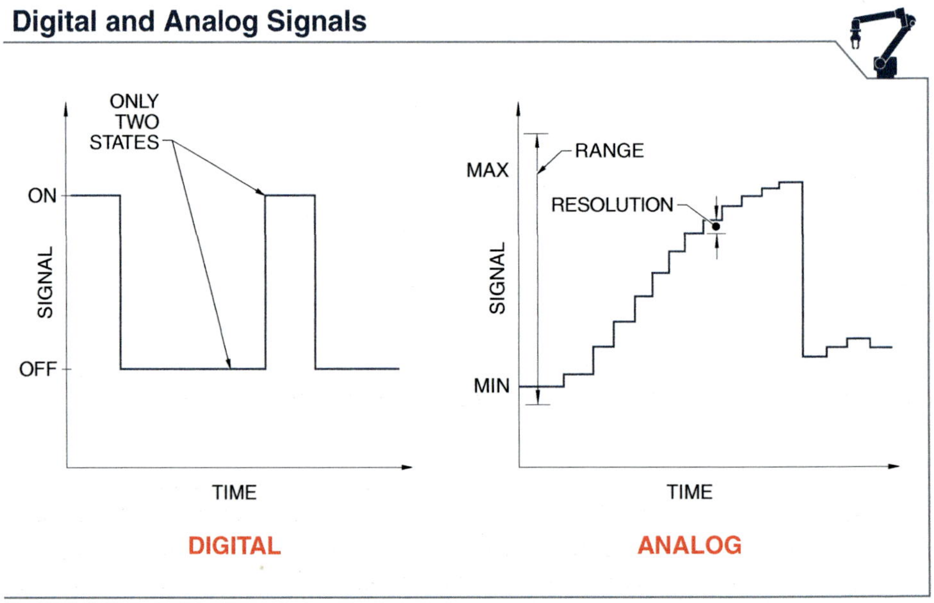

Figure 9-1. Digital signals have distinct ON and OFF states. Analog signals may vary within a range but have a certain minimum incremental step, called resolution.

Robot I/O

Once a robot is fitted with an end effector, there must be some way to control it. *Robot I/O* is the set of digital inputs and outputs between a controller and the robot end effector. FANUC provides eight robot inputs and eight robot outputs. The outputs generally operate solenoids or relays that enable the end effector to interact with other objects, such as closing grippers on an object. Sensors on the end effector provide feedback as robot inputs, such as confirming that the object has been properly grasped and is held firmly in the end effector.

The physical connections for the robot I/O from the end effector to the controller are made in multiple stages. Electrical wires and pneumatic lines to and from an end effector are routed along (or fed through) the robot's upper arm and to its back to an area near the Axis 3 servomotor. **See Figure 9-2.** Here, there is room to mount pneumatic valves and solenoids.

Robot I/O Connections

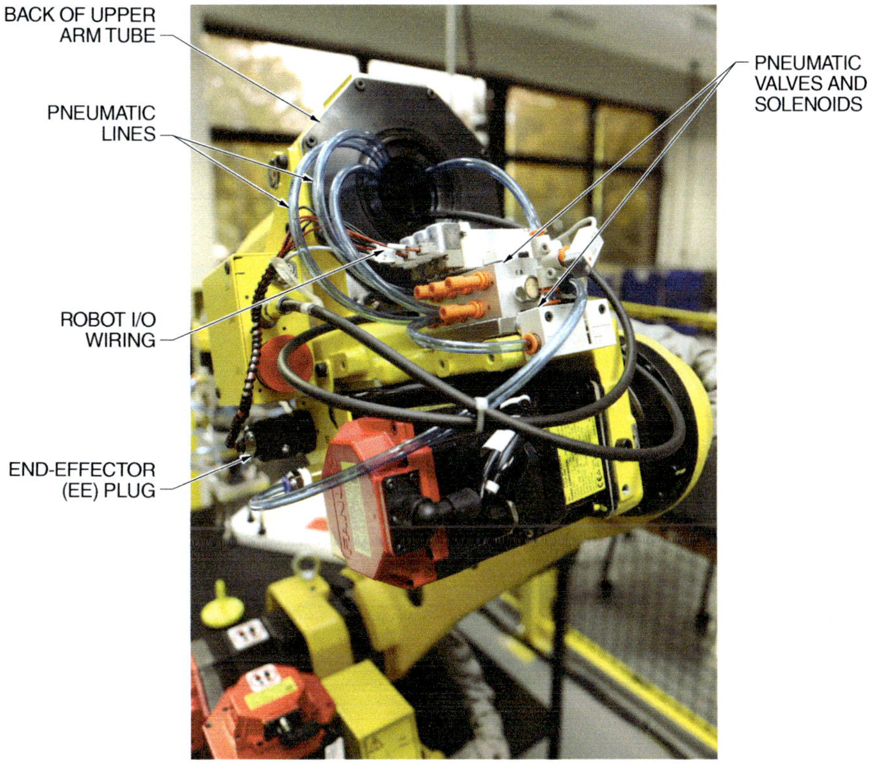

Figure 9-2. Many robot I/O connections are made at the back of the robot's upper arm, near the servomotor for Axis 3. Here, there is space for pneumatic valves and solenoids.

The robot I/O wiring is then collected into a single end effector (EE) plug. This plug has 24 numbered pins, but these numbers do not correspond directly to robot I/O numbers. (Documentation is needed in order to wire the inputs and outputs to the correct pins.) The plug mates with a connector on the side of the Axis 3 joint, which connects to the robot's internal wiring. **See Figure 9-3.** This wiring goes down to the robot base, along with all other robot wiring, where there are connectors for interfacing with the controller.

End-Effector (EE) Plug

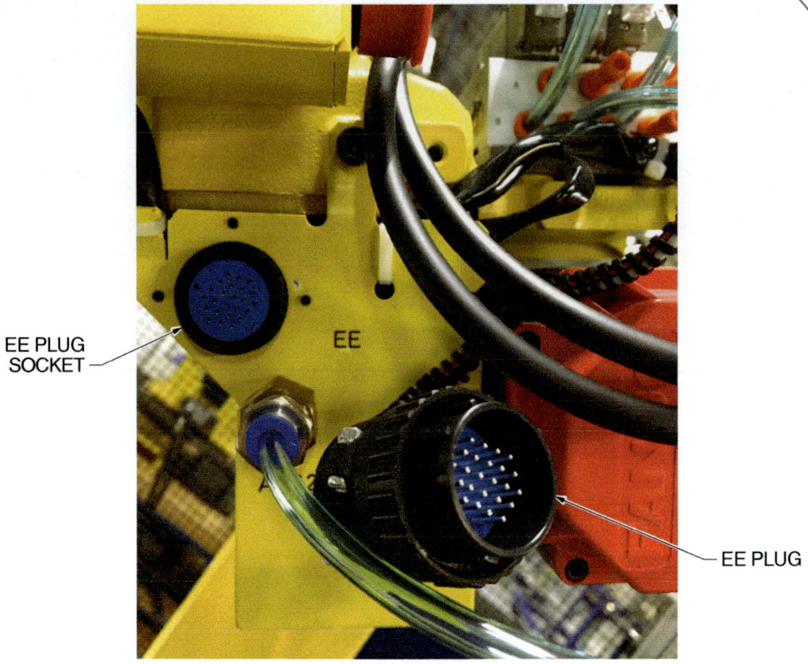

Figure 9-3. The electrical wiring for the robot I/O is collected into the end-effector (EE) plug.

Each robot input or output must be initialized in the controller. To initialize an input or output, a comment is placed in the desired input or output item detail stating what it will be used for. **See the *Initializing Complementary Robot I/Os* Procedure.**

In some cases, outputs must be linked in a way that does not allow them to have the same value at the same time. FANUC calls these outputs complementary. *Complementary outputs* are two outputs that must always have opposite statuses (ON or OFF). For example, the output for opening a gripper is complementary to the output for closing the gripper because they cannot both be ON or both be OFF at the same time. After establishing the complementary outputs, when one output is turned ON, the complementary output automatically turns OFF. This prevents two electrical solenoids from engaging at the same time. In order for the complementary option to work, the robot must be power cycled after the outputs are set up.

Collaborative Robots

A collaborative robot (cobot) is designed to operate alongside a human operator or without a fence. Cobots are equipped with hardware and software that slow the robot speed when people are sensed near it. Some cobots have touch sensors in their housings that are designed to immediately stop the robot if they are touched. The robot joint mechanisms are also cushioned to prevent injury if the arm comes into contact with a person. FANUC cobots are differentiated by their green exterior, so that it is clear to personnel that the cobots have these safety features.

The primary purpose for a cobot is to function where people must be present to operate or inspect other equipment. Though this style of robot is generally simple to program and can be installed in areas without fencing, it operates slower than a standard robot. In addition, a cobot must have additional safety software and hardware to ensure operators are safe working near it.

PROCEDURE

Initializing Complementary Robot I/Os

Set Up Robot Output 1

1. Press I/O (near the bottom of the TP).
2. Press F1 [TYPE].
3. Highlight Robot and press ENTER to show the I/O Robot Out screen.

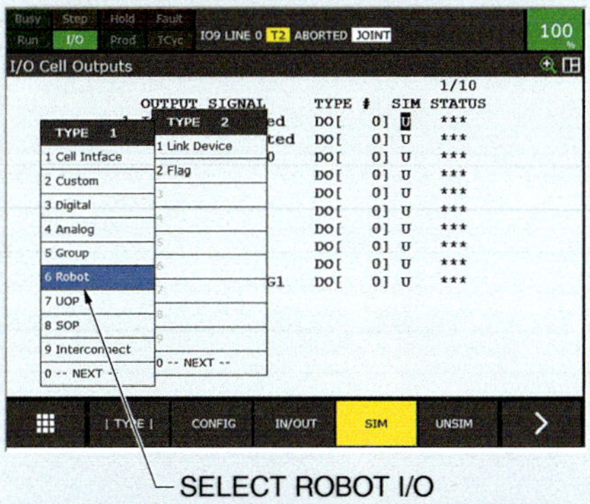

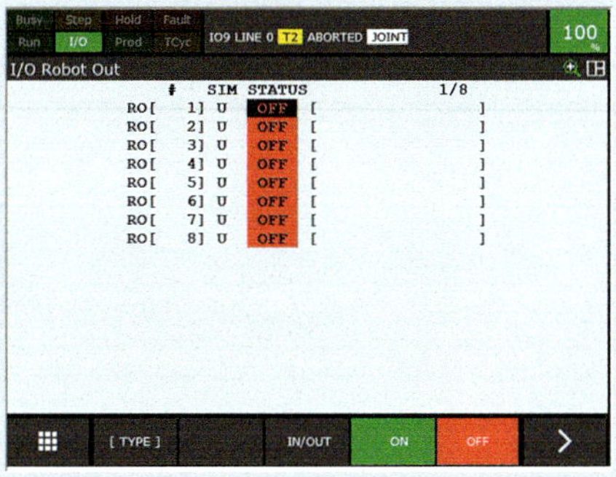

4. The first item in the outputs list is highlighted. Press NEXT and then press F4 DETAIL.
5. Highlight the comment field. Press ENTER and type in the desired comment to identify the output. In this example, it is “OpenGripper”.
6. Highlight FALSE in line 3. Press F4 TRUE. The screen shows Power OFF then ON to enable changes.

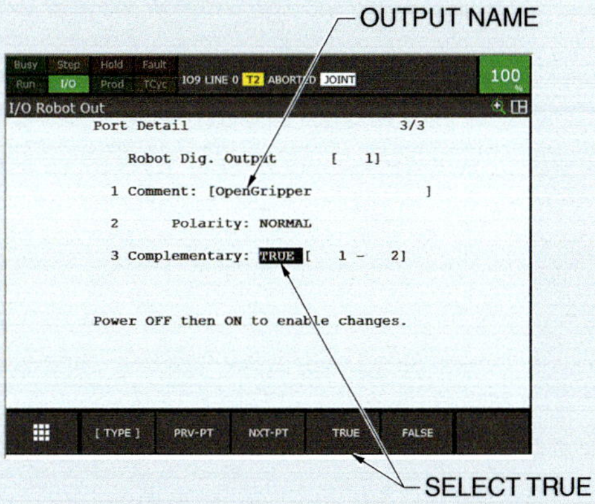

7. Press PREV.

Set Up Robot Output 2

⑧ Highlight the second output and press [F4] DETAIL.

⑨ Highlight the comment field. Press [ENTER] and type in the desired comment. In this example, it is "CloseGripper".

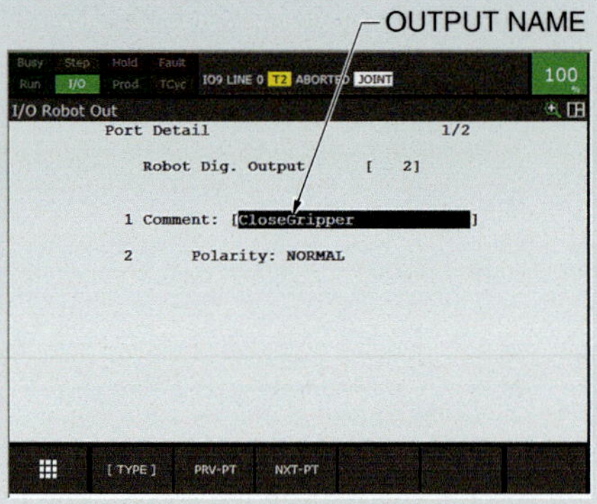

Cycle the Robot Power

⑩ Press [FCTN].

⑪ Highlight -- NEXT -- and press [ENTER].

⑫ Highlight CYCLE POWER and press [ENTER].

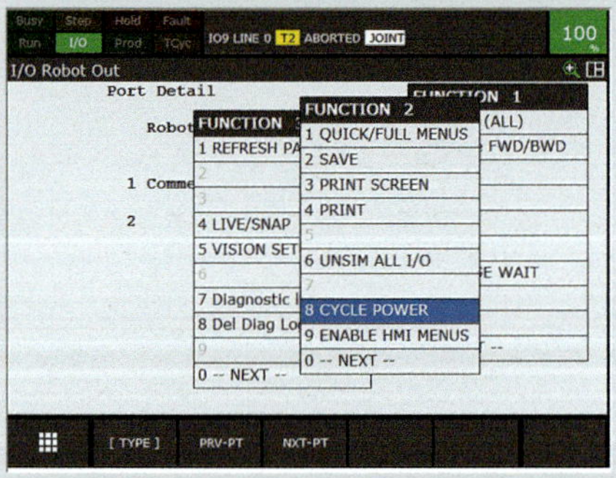

⑬ The screen shows This will cycle power. Are you sure? Select YES.

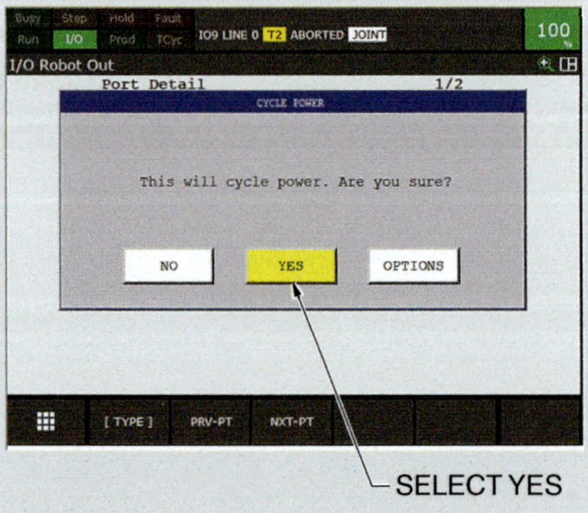

Testing the Complementary Outputs

⑭ Return to the `I/O Robot Out` screen.

⑮ Highlight the status of Output 1.

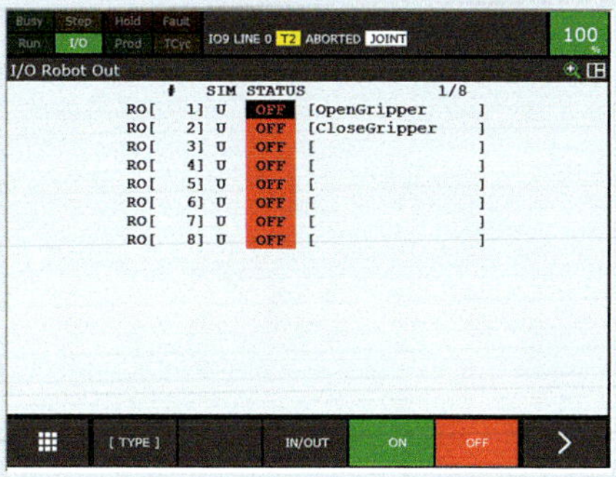

⑯ Alternate pressing F4 `ON` and F5 `OFF` while noting the statuses of Outputs 1 and 2.

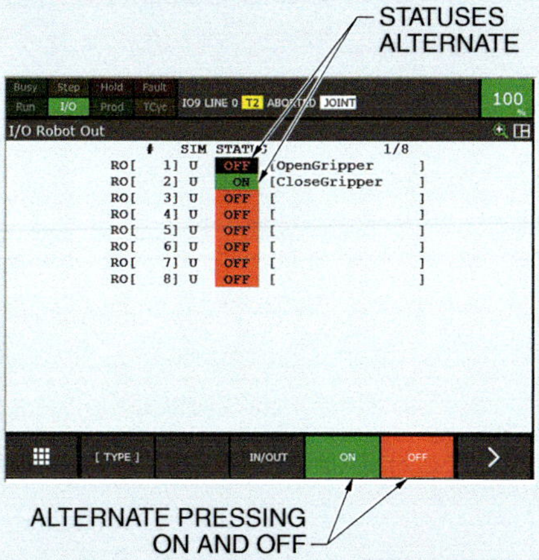

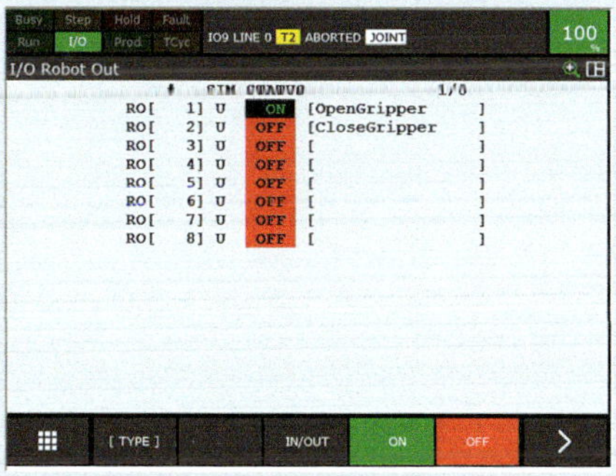

⑰ Highlight the status of Output 2 and repeat Step 16.

When the robot end effector is directly above the object to be picked up, a linear movement at a reduced speed is used to bring the gripper into position. It is also important to align and square the end effector with the object, which allows the object to be grabbed without the object moving. **See Figure 9-4.**

End-Effector Pneumatic Operation

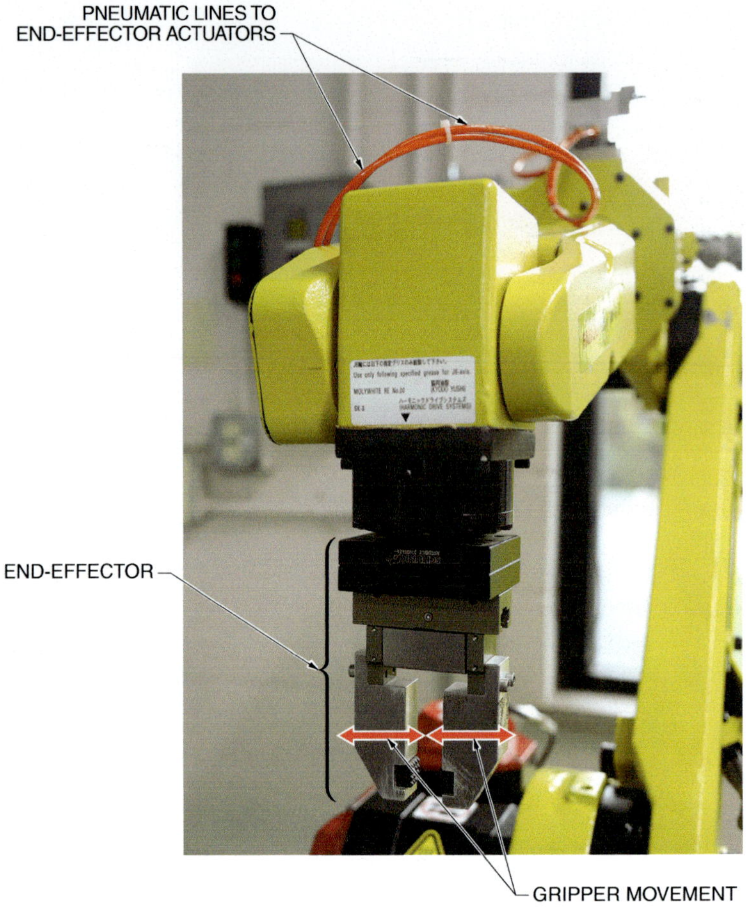

Figure 9-4. Pneumatic systems are used to actuate moving parts on an end effector. The air pressure is used to move air cylinders.

Once the gripper end effector is in position, a robot output instruction is used to close the gripper on the object. **See the *Gripping an Object with the End Effector* Procedure.** However, to ensure the object is gripped properly, a Wait instruction must be placed directly after the close-gripper instruction. This is because the robot can read the program and move more quickly than the gripper can move mechanically. To close the gripper, the program sends an electrical signal to the "close" solenoid. The solenoid energizes, causing a pneumatic spool to shift position. Air then flows to move a piston and close the gripper on the object. These actions take several milliseconds to complete, which may be slower than the robot reading and executing the code. Without the delay, the robot would continue executing its program before the object is gripped properly.

The Wait instruction can be a simple time delay in seconds. However, in this example, the end effector is equipped with a "part-present" sensor that signals ON when the gripper senses the object. This robot input can be used to control the Wait instruction in the most efficient way.

PROCEDURE

Gripping an Object with the End Effector

Set Up Robot Input 1

① Press I/O (near the bottom of the TP).

② Press F1 `[TYPE]`.

③ Select `Robot` and press ENTER to show the `I/O Robot Out` screen.

④ Press F3 `IN/OUT` to toggle to the `I/O Robot In` screen.

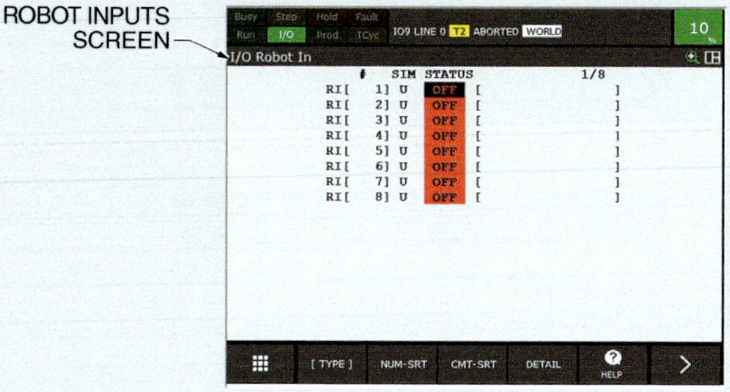

⑤ The first item in the inputs list is highlighted. Press NEXT and then press F4 `DETAIL`.

⑥ Highlight the comment field. Press ENTER and type in the desired comment to identify the input. In this example, it is "GripperClosed".

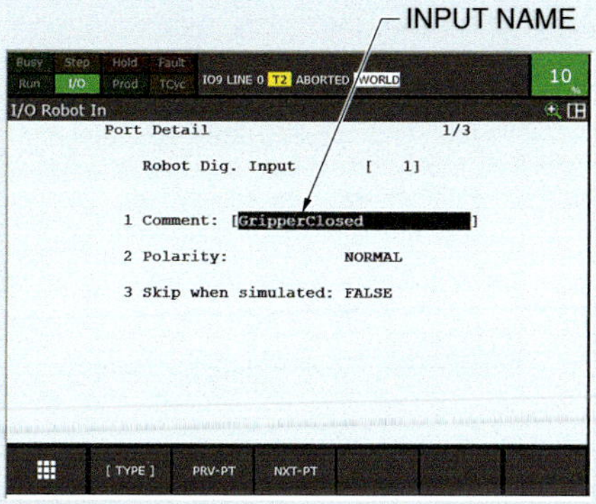

Add Robot Output Instruction

⑦ In the program, highlight the appropriate blank line.

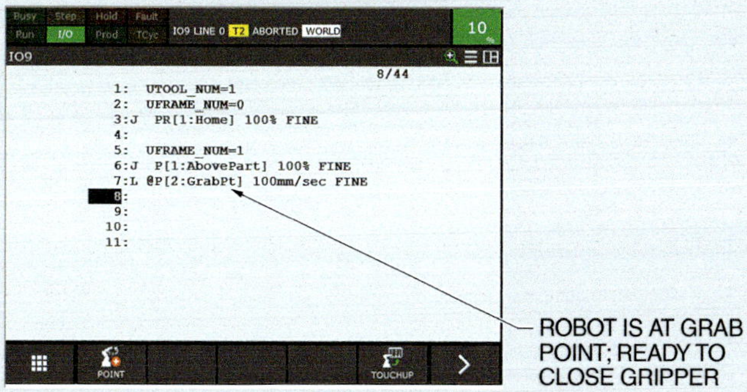

⑧ Press F1 `[INST]`.

⑨ Highlight `I/O` and press ENTER.

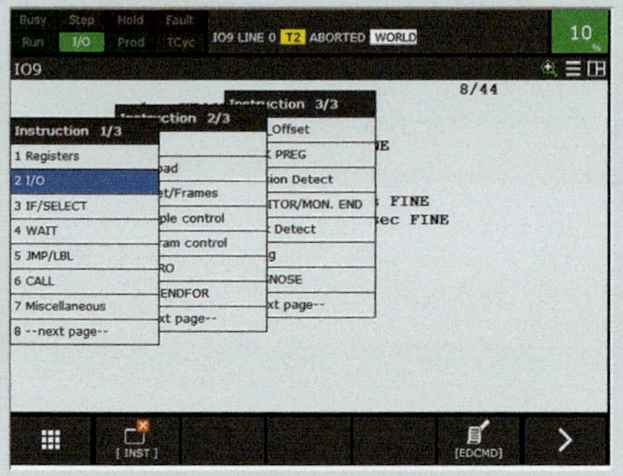

⑩ Highlight `RO[ ]=...` and press ENTER.

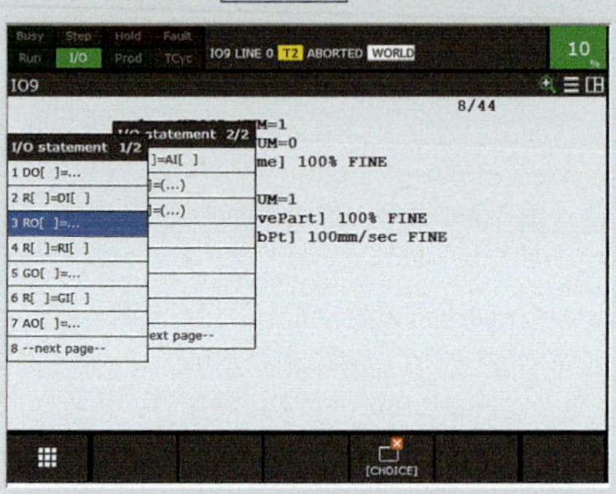

⑪ The space between the brackets is highlighted. Type the number of the desired robot I/O. In this example, it is 2 for closing the end effector. Press ENTER.

⑫ The space after the equal sign is highlighted, and a menu appears. Highlight `On` and press ENTER.

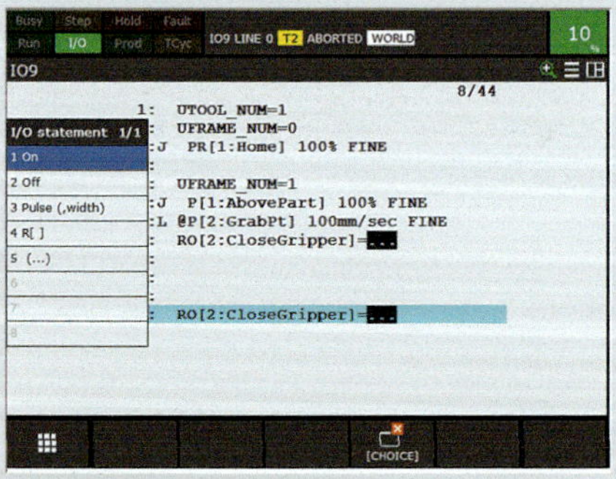

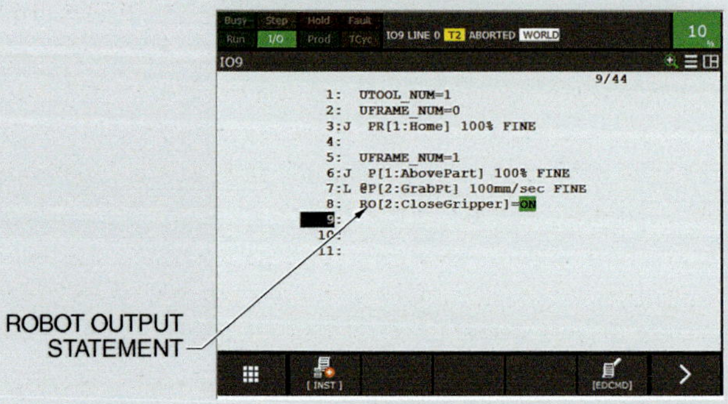

Add a Wait Instruction with Robot Input

(13) Highlight the blank line directly after the instruction to close the gripper.

(14) Press F1 [INST].

(15) Highlight WAIT and press ENTER.

(16) Highlight WAIT ...=... and press ENTER.

(17) A menu appears. Highlight RI[ ] and press ENTER.

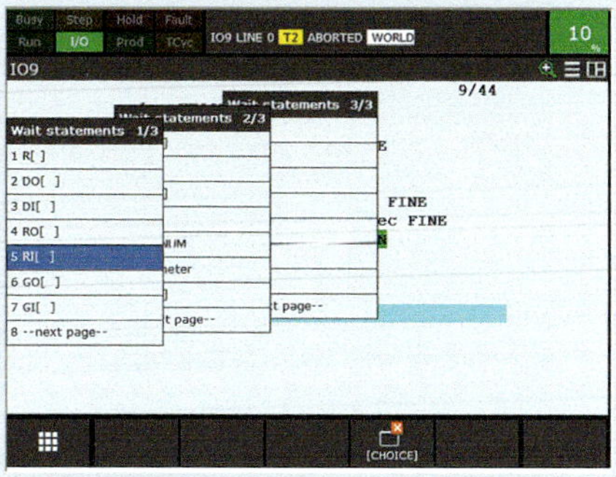

(18) The space between the brackets is highlighted. Type the number of the desired robot input. In this example, it is 1. Press ENTER.

(19) A menu appears. Highlight On and press ENTER.

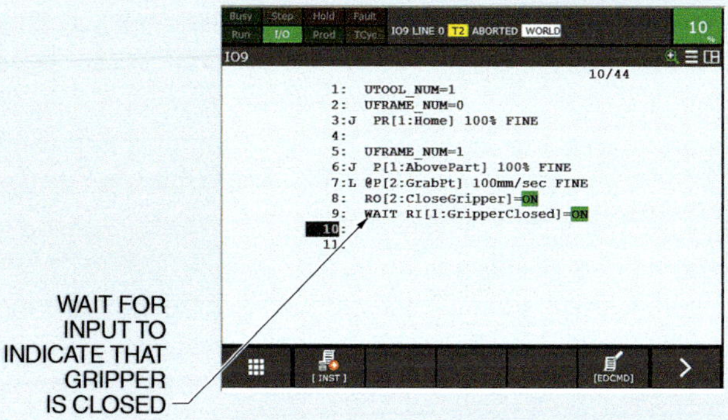

Digital I/O

When robots interact with other moving devices or equipment in the cell, the programmer must incorporate information on these items to both coordinate tasks and avoid damage. For example, if a robot must place an object into a machining center, the program must wait until the cutting tool is out of the way and the previously machined part has been removed. Or the robot may have to wait for a conveyor belt to bring an object to the proper position before the object can be picked up. Sensors are used to detect these conditions and provide this information to the robot controller, where this information is incorporated into the program.

Digital I/O is the set of digital inputs and outputs between a controller and other cell equipment. Digital input signals are commonly generated by switches that detect a moving object, a nearby object, or an object in position. Digital outputs are often used to open/close doors, turn signal lights ON/OFF, or activate/deactivate processes.

The wiring from digital input and output devices is connected to a PLC, which is then interfaced to the robot controller via an Ethernet connection. **See Figure 9-5.** Set-up procedures are needed in order to link the physical I/O devices to numbers in the controller software.

I/O PLC Interface

Figure 9-5. Input and output devices are connected to a controller through an external PLC.

Consider a robot that loads and unloads parts for a machining center. The programmer must prevent the robot from entering the machining center unless the door is open. To ensure the door is open, a `WAIT DI[ ]=ON` instruction is entered into the robot program. This pauses the program at this line until this digital input is turned on. When the machining center's door opens, it activates a limit switch, which sends the digital input to the controller, and the robot program continues.

As with the robot inputs and outputs, digital inputs and outputs need to be initialized. This is done in the same manner as robot inputs and outputs. **See the *Initializing Digital I/Os* Procedure.** Digital outputs can also be set up as complementary.

PROCEDURE

Initializing Digital I/Os

1. Press I/O.
2. Press F1 `[TYPE]`.
3. Highlight `Digital` and press ENTER to show the I/O Digital Out screen.
4. Press F3 `IN/OUT` to show the I/O Digital In screen.

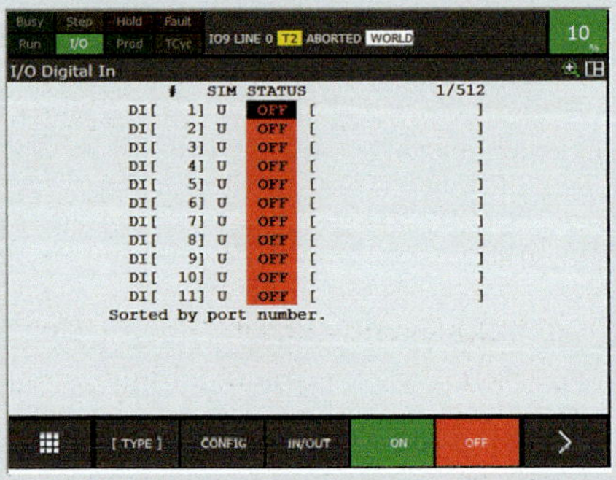

5. Highlight the desired input number. Press NEXT and then press F4 `DETAIL`.

⑥ Highlight the comment field. Press ENTER and type in the desired comment. In this example, it is "DoorOpen".

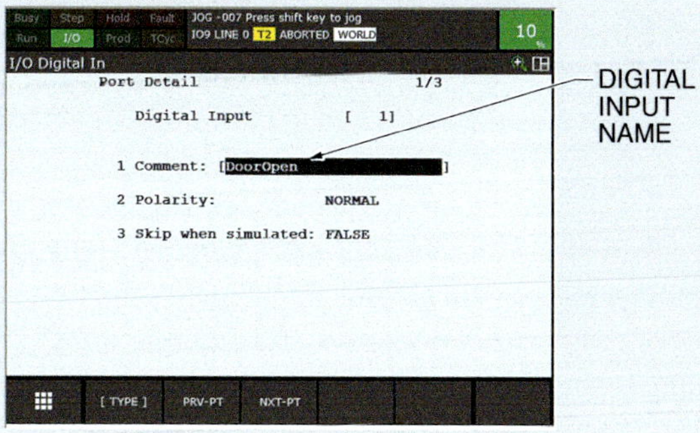

The digital input has now been initialized and will provide a descriptive comment next to the digital input in the program. The next step is to place the digital input command and Wait instruction into the robot program. **See the *Waiting for a Digital Input* Procedure.**

PROCEDURE

Waiting for a Digital Input

① Highlight a blank line.
② Press F1 [INST].
③ Highlight WAIT and press ENTER.
④ Highlight WAIT ...=... and press ENTER.
⑤ Highlight DI[ ] and press ENTER.
⑥ The space between the brackets is highlighted. Type the number of the desired digital I/O. In this example, it is 1. Press ENTER.
⑦ The space after the equal sign is highlighted, and a menu appears. Highlight On and press ENTER.

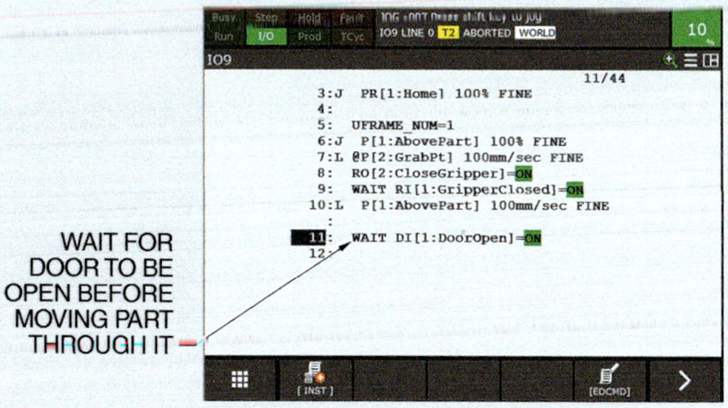

Group I/O

Group I/O is a set of digital inputs and outputs that are assigned to be activated (turned ON and OFF) as a group. For instance, if the machining center's door is open and ready to load a product, several inputs may need to be activated at once. This can be done in two ways. The programmer can enter several lines in the program to turn on each input individually or create a group with one program line activating all the inputs. A group program line saves programming time and reduces the length of the program.

Analog I/O

The last type of I/O is analog. *Analog I/O* is the set of analog inputs and outputs between a controller and other cell equipment. Analog inputs and outputs are connected physically, initialized, and inserted into programs in the same way as digital inputs and outputs, except that their values can be more than just ON or OFF.

Analog outputs are most commonly used to control the speed of a machine tool in the cell. For example, a grinding wheel inside a manufacturing cell must rotate at a speed so that the outer edge has a specific surface speed. **See Figure 9-6.** The speed is important for shaping parts effectively. The robot outputs an analog signal to control the speed. As it is used, the grinding wheel wears down, reducing its diameter and the surface speed at its edge. The robot program keeps track of the number of part cycles and the expected wear on the grinding wheel. The program accounts for the change in diameter by gradually increasing the rotation speed (via the analog output) after a set number of part cycles. This keeps the wheel edge at the optimum surface speed.

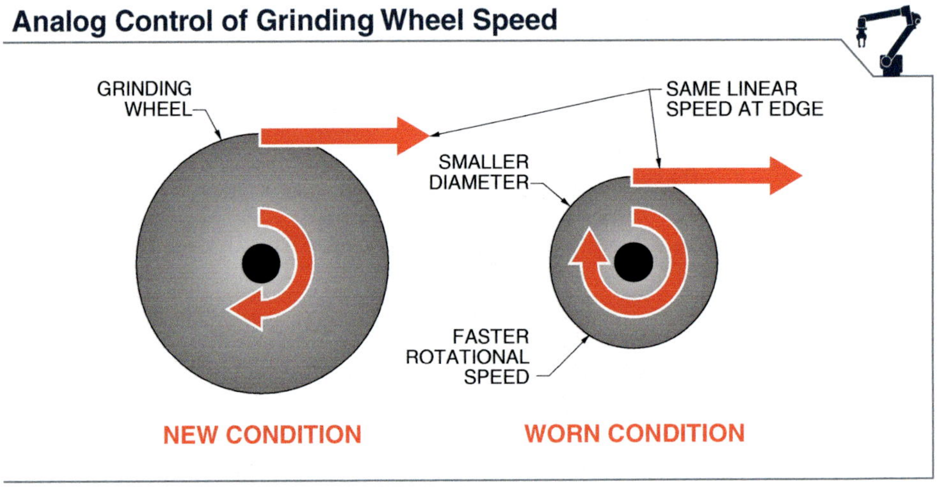

Figure 9-6. Analog output signals are commonly used to control the speed of machine tools, such as grinding wheels. In this case, the wheel speed must be adjusted as the wheel diameter decreases from wear.

When the grinding wheel is replaced, a reset is needed. The reset is manually activated and changes the analog output value back to the starting value for a full-diameter wheel.

A complication to this program is that the force the robot uses to push the part into the grinding wheel affects how much material is removed from the part and how fast the grinding wheel wears. The solution is to allow the robot to push the part into the wheel until a certain amount of force is measured.

There are at least two ways to accomplish this. The first is to put the grinding wheel on a floating carriage that is equipped with a pneumatic cylinder. The air pressure in the cylinder corresponds to the force the part is putting on the grinding wheel. The pressure can be measured with a sensor and fed back to the controller as an analog input, where it can be used by the program to adjust robot movement into the grinder. However, this method has flaws. For example, the pressure will vary as the air pressure changes.

A second method is to use the FANUC Force Sensing add-on software and hardware. This software and hardware work together to provide the desired pressure as the grinding wheel changes diameter. Also, if the robot must rotate the part, the Force Sensing add on can adapt to this movement to maintain constant force on the grinding wheel.

SECTION 9.2 INTEGRATING END EFFECTORS

When a company orders a robot from the manufacturer, they receive what is referred to as a stripped-down robot. A *stripped-down robot* is a robot that has not yet been equipped with an end effector. **See Figure 9-7.** With so many different requirements and options for end effectors, it is not practical for robot manufacturers to include end effectors with robot models. In fact, FANUC does not produce end effectors. Instead, other companies are relied on to source compatible end-effector solutions or components that can be used to make custom end effectors. Robot customers may do this themselves or hire an integrator. An *integrator* is a company that designs, installs, and programs complete robot systems for industrial applications.

Stripped-Down Robots

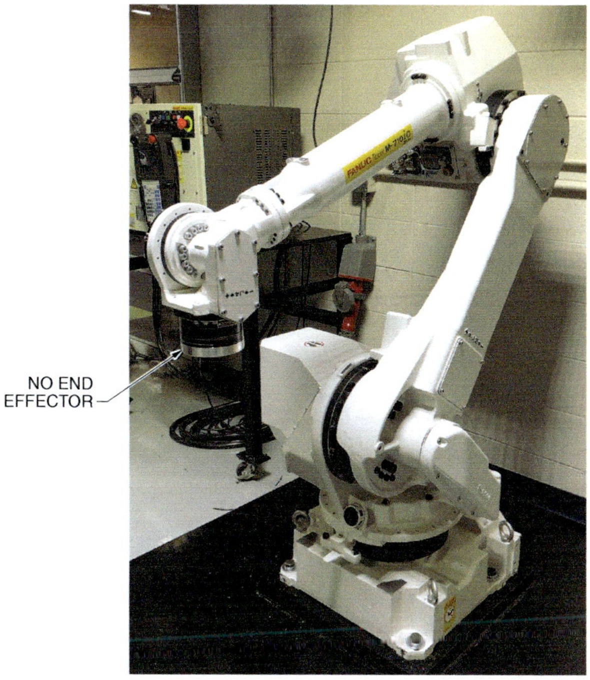

Figure 9-7. A stripped-down robot does not include an end effector.

Choosing End Effectors

Possible end-effector shapes and sizes are as numerous as there are products in the world. Some end effectors are large and can pick up multiple products. Others are very small and designed for delicate assemblies. Each is custom designed to execute a particular robot task. Several factors must be considered when choosing or designing an end effector.

End-Effector Rigidity. The end effector must be rigid enough to allow repeatability when picking and placing a part. *Repeatability* is the ability of a robot to return to the same position many times within a very narrow tolerance. Robots are commonly capable of repeatability to within 0.0002″. The end effector must stay stable and not shift from one pick to the next or repeatability would not be maintained. There is typically very little tolerance in fixtures for poorly placed parts. **See Figure 9-8.**

Repeatability

Figure 9-8. Repeatability is the ability of a robot to return to a precise position over and over again. This is especially important for picking and placing objects into fixtures.

End-Effector Types. There are many possible methods for end effectors to handle products. Sometimes products that are very delicate, oddly shaped, or made from unusual materials require unique end effectors. Most robots, however, are equipped with either pneumatic or mechanical-gripping-type end effectors. These two types are suitable for the vast majority of products handled by robots.

Pneumatic end effectors are commonly used for thin, sheet-like, lightweight, or more delicate products. The most common type is the venturi system. **See Figure 9-9.** This system directs a compressed air supply through a restriction that speeds up the air, lowering its pressure. This low pressure pulls up air from a suction cup that, when pressed against a product surface, has sufficient force to hold the product. Another pneumatic system uses a vacuum pump to draw air from suction cups, which creates a vacuum that can hold a product. The venturi system is more common, however, since it is usually more reliable and less expensive than a vacuum-pump system.

Pneumatic End Effectors

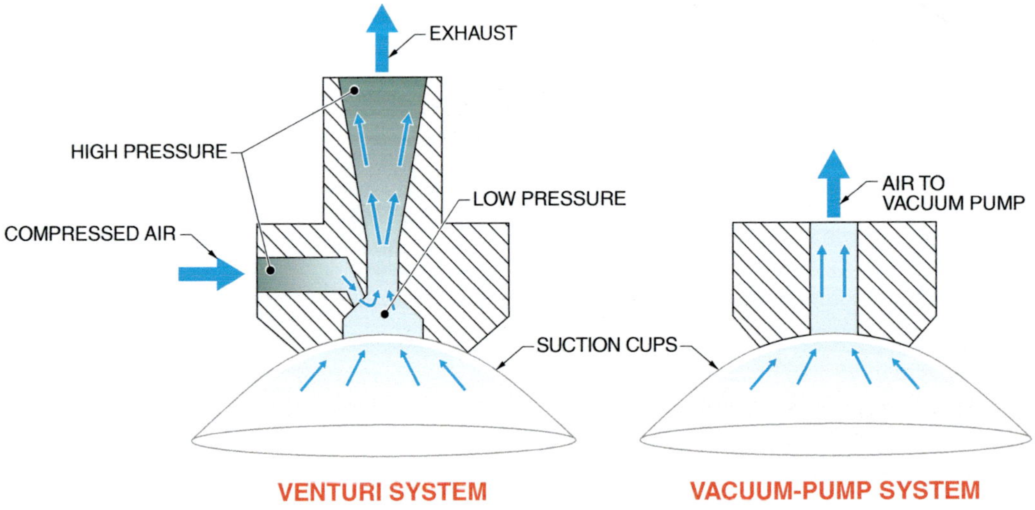

Figure 9-9. A venturi system uses compressed air to create a gripping suction. A vacuum-pump system uses a vacuum pump to create a gripping suction.

Intrinsically Safe Teach Pendants

A standard teach pendant (TP) cannot be used in a hazardous atmosphere, which can be created by dust or vapor from industrial processes. Such environments are prone to explosion if exposed to an ignition source, such as a small spark from conventional electronics. Therefore, all equipment in hazardous atmospheres must be marked as conforming to standards for intrinsic safety.

Intrinsically safe electronics use various enclosure and/or encapsulation methods to prevent the production of sufficient heat or sparks to ignite an explosive atmosphere. Many manufacturers of electronic devices that are used in industry also produce alternate versions of their products that are intrinsically safe. For example, FANUC provides the option for an intrinsically safe TP.

Explosive atmospheres are somewhat common in industrial settings. Engineers or technicians specifying the proper robot for a job should be aware of these hazardous conditions and specify the correct and safe equipment to use in these areas.

Grasping with mechanical finger-like grippers is another way to handle a product. These are well-suited for parts, either large or small, made from hard materials. Grippers with numerous variations in size, number of fingers, and gripping force are available from end-effector manufacturers. The fingers are actuated with small, pneumatic cylinders. However, grippers are still considered mechanical end effectors rather than pneumatic.

Grippers often come with relatively small fingers with threaded holes, which allow custom finger extensions to be attached. **See Figure 9-10.** This way, integrators can design and make gripper fingers that are perfectly matched to a product. Fingers can also be designed to either grip the outside of a product by pressing in or, if the product has a hole, to grip the inside of a product by pushing out.

Grippers

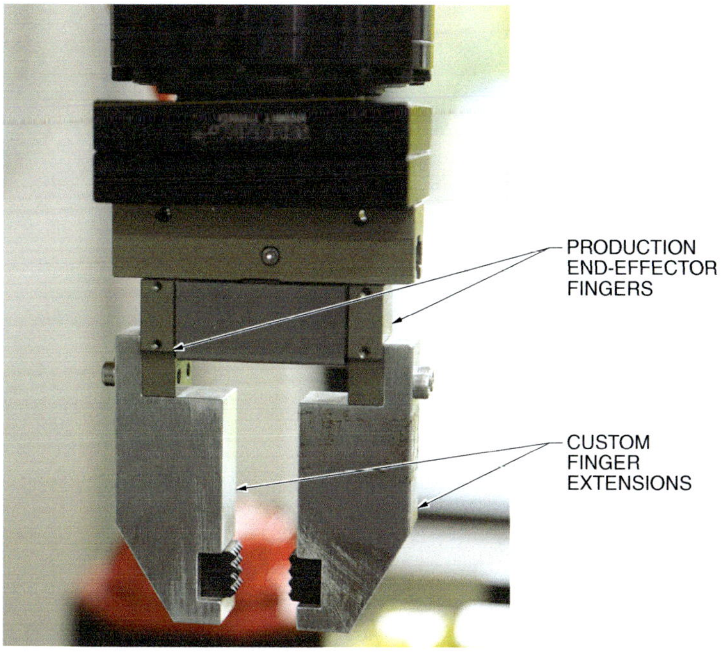

Figure 9-10. Many designs of grippers allow custom-machined extensions to be added to the fingers. This allows the gripper to be matched perfectly with the parts to be handled.

Combination End Effectors. The next consideration in end-effector development is how many types of products the end effector must handle. Manufacturers often have several product variations, each needing to be picked and placed accurately. Or a robot may need to handle more than one type of item for a task. For example, a robot that picks product and places it on a skid may also need to be able to handle a skid sheet. (A skid sheet is a cardboard sheet the size of the skid that is placed between each layer of products.) That robot may also need to pick up and place an empty skid. Thus, the same end effector must be able to handle three totally different items.

A possible solution is a combination end effector. A *combination end effector* is an end effector composed of multiple types of tooling or grippers. In this example, the end effector could have mechanical fingers to hold the product, suction cups to hold the skid sheet, and a very large second set of fingers to hold the skid.

Attaching End Effectors

If the design of an end effector satisfies all the tasks required of it, it can have a rigid attachment. A *rigid attachment* is an attachment of an end effector to a robot that requires significant downtime to remove. Maintenance must be called in to disconnect the pneumatics and powered components and unbolt the end effector from the robot. This process can be quite lengthy and is normally only done if the end effector is badly damaged and requires extensive maintenance.

Another way to connect an end effector is with a tool changer. **See Figure 9-11.** A *tool changer* is an electronically controlled module installed between a robot and an end effector that enables the robot to automatically release one end effector and attach another. The tool changer requires electrical signal connections to the controller and can be commanded to release or attach an end effector through the teach pendant (TP) via robot I/O or digital I/O. This makes the robot versatile, but the equipment can be costly.

Tool Changers

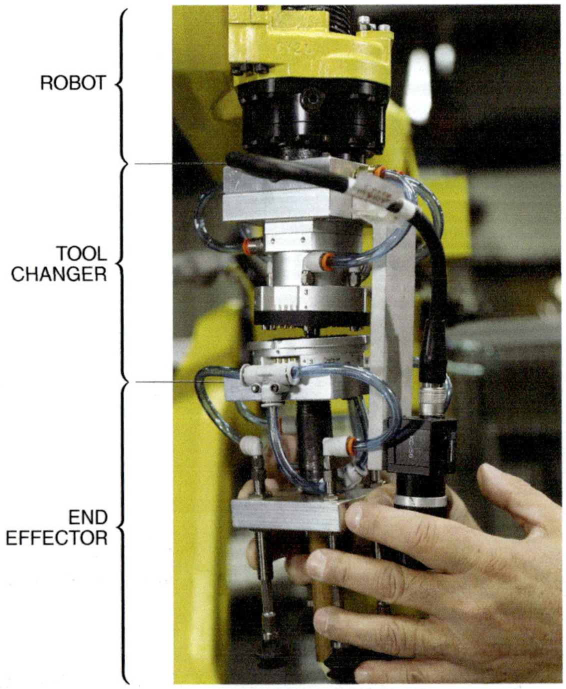

Figure 9-11. A tool changer can be installed between a robot and an end effector. It can then be commanded from the TP to release the end effector and allow another one to be attached quickly.

For example, a technical college may use the same robots to teach both material handling and robotic welding. In this case, the robots can be equipped with automatic tool changers that allow the use of pneumatic end effectors for the morning material-handling class and the ability to change to simulated welding torches for the afternoon welding class.

Part-Present Devices

Robots running automated processes often do so without an operator present, or one operator may be assigned to watch several automated robot cells. With no one watching a robot, if a failure occurs, the robot should be signaled to fault and stop operation automatically. This is important in order to prevent damage to parts or equipment.

A common reason for a fault-and-stop-operation program is when a robot fails to pick a part correctly. This includes a part that is totally missed or one that slips partially or fully out of the end effector. A part-present device on the end effector detects such failures. A *part-present device* is an electrical switch on an end effector that senses the presence of a picked part and indicates if it is properly held. Different types of switches may be used depending on the end-effector design, but all have the same purpose. **See Figure 9-12.**

Part-Present Devices

Figure 9-12. A part-present device indicates to the controller if a gripper is properly holding a part. The device is installed on the end effector in such a way that it is only activated when a part is held properly.

A *limit switch* is an ON/OFF switch that is activated when something pushes on its lever arm. The arm must be positioned on the end effector in such a way that only a proper grip on the part will push the lever arm.

A *proximity switch* is an ON/OFF switch that senses the presence of a nearby object without physical contact. There are different types of proximity switches using different technologies. Generally, they emit either an electromagnetic field or an infrared light and then detect changes in these caused by nearby objects. Some types are suitable for all part materials, and some detect only metal. A proximity switch must be carefully mounted on an end effector to prevent the end effector itself from activating the switch. Proximity switches have the advantage of having no moving parts, so they are not easily damaged.

A *pressure switch* is an ON/OFF switch that senses a change in pressure through the use of a movable diaphragm. A pressure switch can only be used with pneumatic (suction-type) end effectors. When the end effector applies suction at first, the pneumatic system is open, and the pressure change is small. When the end effector's suction cup then contacts the part, the pneumatic system is closed, and the pressure switch senses the vacuum (negative pressure) and activates. The switch must be properly configured to work in a negative pressure range.

All of these switches serve the same purpose. When activated by a part, the switch closes a circuit and sends a digital ON (also known as a HIGH or TRUE) signal to the controller. When the end effector does not hold a part properly, the switch remains open, and the controller sees the signal as a digital OFF (also known as a LOW or FALSE) signal. If the signal remains OFF when the controller is expecting an ON (when the end effector is commanded to pick a part), then the robot program should fault. Conditional statements can be added to the robot program to redirect the robot to an error-handling subprogram in such an event. **See Figure 9-13.** The subprogram stops the robot and may display a message to the operator with the type of fault.

Example of Error Handling with Part-Present Switch Input

```
:J  P[1:AbovePart] 100% FINE
:L  P[2:GrabPt] 100mm/sec FINE
:   RO[2:CloseGripper]=ON
:   WAIT    .50(sec)
:   IF RI[1:PartPresent]=OFF, CALL ERR_PARTGRAB
```

IF PART IS NOT PROPERLY GRABBED,
THEN CALL AN ERROR SUBPROGRAM

Figure 9-13. The input signal from a part-present device can be used in a program to handle picking errors. For example, if the signal remains OFF, the robot is redirected to an error-handling subprogram.

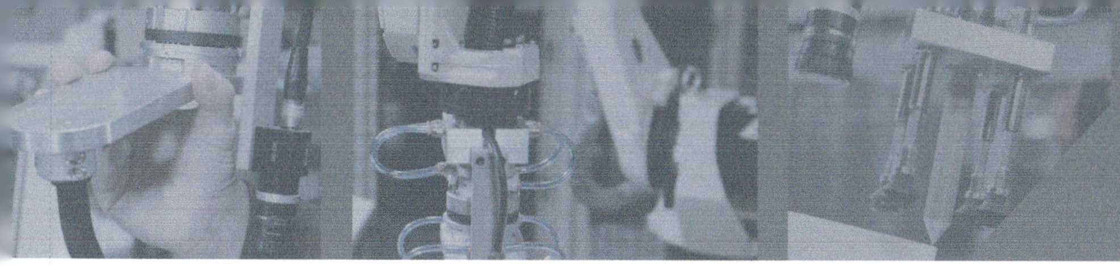

Review 9

Name ______________________ Date ______________

SECTION 9.1 UTILIZING INPUTS/OUTPUTS

______________ **1.** ___ signals typically come from sensors reacting to some aspect of the environment, such as a door switch closing.
A. Output
B. Input
C. Analog
D. Offset

______________ **2.** A(n) ___ signal is an electrical signal that indicates only ON or OFF (or HIGH or LOW) states.

______________ **3.** ___ outputs are two outputs that must always have opposite statuses (ON or OFF).
A. Digital
B. Group
C. Collaborative
D. Complementary

______________ **4.** In order for the complementary option to work, the robot must be ___ after the outputs are set up.

______________ **5.** Digital input signals are commonly generated by ___ that detect a moving object, a nearby object, or an object in position.

______________ **6.** Like robot outputs, ___ outputs can also be set up as complementary.
A. analog
B. digital
C. group
D. primary

7. Why are inputs and outputs necessary for interacting with other cell equipment, such as a machining center?

8. What are common uses of robot inputs and outputs?

9. Why might a Wait instruction be needed after programming a gripper to close on an object?

10. What is the purpose of a group I/O?

11. How are analog outputs most commonly used?

SECTION 9.2 INTEGRATING END EFFECTORS

_______________ 12. A stripped-down robot is a robot that has not yet been equipped with a(n) ___.

_______________ 13. ___ is the ability of a robot to return to the same positions many times within a very narrow tolerance.

_______________ 14. The most common type of pneumatic end effector is a ___ system.

A. finger
B. vacuum
C. venturi
D. compression

_______________ 15. The fingers of a gripper are actuated ___.

A. pneumatically
B. with a motor
C. with a vacuum
D. manually

_______________ 16. A(n) ___ end effector is an end effector composed of multiple types of tooling or grippers.

_______________ 17. A ___ attachment is an attachment of an end effector to a robot that requires significant downtime to remove.

A. tool changer
B. rigid
C. part-present
D. combination

_______________ **18.** A ___ switch is an ON/OFF switch that is activated when something pushes on its lever arm.

A. proximity
B. lever
C. pressure
D. limit

_______________ **19.** A(n) ___ switch can only be used with pneumatic (suction-type) end effectors.

20. Why are robots not equipped with end effectors from the factory?

21. What is an integrator?

22. How are gripper fingers customized to match the product being picked?

23. What is a tool changer?

24. What is the purpose of a part-present device?

25. How can a controller use the feedback from a part-present device?

9A
Lab Exercise

Lab Objective

Identify the inputs and outputs that are available on a robot.

Equipment Required

- Robot with a variety of connected robot, digital, and/or analog inputs and/or outputs

Lab Procedure

1. Explore the I/O screens to see what inputs and outputs have been initialized on the robot.
2. Document every available input and output in a spreadsheet format. Organize them by type. Include designations (e.g. "RO1"), descriptions, and whether any are complementary.

9B
Lab Exercise

Lab Objectives

- Research a compatible end effector and its mechanical, electrical, and/or pneumatic requirements.
- Prepare a functional drawing showing the connections of the robot-end-effector system.

Equipment Required

- Robot with end effector
- Robot documentation
- End-effector documentation

Lab Procedure

1. Look up mechanical specifications and drawings of the robot, which should be on either a USB drive or a "mechanical" information CD supplied by FANUC.
2. Determine end-effector-related information about the robot:
   - end-of-arm bolt pattern
   - number of fasteners
   - fastener size and thread
   - location of end-effector (EE) plug
3. Look up the specifications and drawings of the end effector. If this documentation is not already available, it may be found on the manufacturer's website.
4. Determine end-effector operating and connection information:
   - type of end effector (gripper, pneumatic venturi, pneumatic vacuum, etc.)
   - method of actuation (compressed air, electricity, etc.)
   - type of actuation activation (solenoid pneumatic valve, electrical relay, etc.)
5. Determine the robot I/O connections required and their pin numbers for the EE plug.
6. Draw a functional diagram of the end effector and its mechanical, electrical, and/or pneumatic connections to the robot. See Example Drawing.

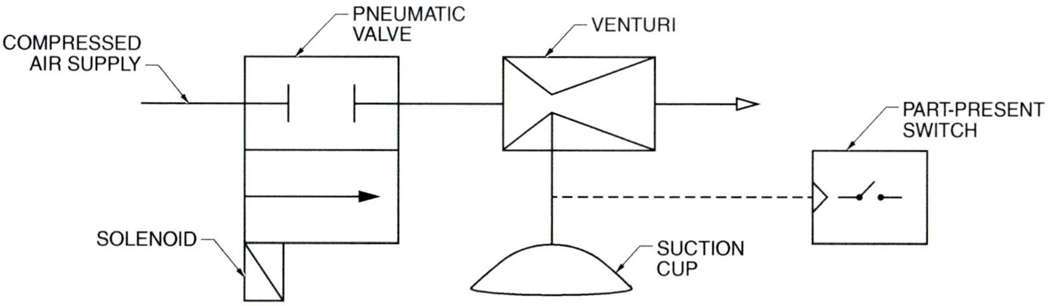

10 Using ROBOGUIDE® Software

For students learning robotics, the ROBOGUIDE software can assist in getting started. The software costs far less than an actual robot and allows unlimited practice time that builds experience and confidence. This software can provide that "robot time" when a robot is not available. The ROBOGUIDE software has many options that may appear somewhat complicated and confusing. However, some guidance on software selection, registration, and some useful features can make getting started much easier.

Objectives

SECTION 10.1 CREATING A ROBOGUIDE CELL

- Describe the process of and options for creating a new, virtual robot cell in ROBOGUIDE.
- Demonstrate how to position and size an end-effector and set a TCP in ROBOGUIDE.
- Demonstrate adding and placing fixtures within a virtual robot cell in ROBOGUIDE.

SECTION 10.2 NAVIGATING ROBOGUIDE

- Describe the basic mouse functions for navigating ROBOGUIDE.
- List common menu commands and toolbar icons in ROBOGUIDE.
- List the differences between the virtual teach pendant in ROBOGUIDE and an actual teach pendant.

SECTION 10.3 PROGRAMMING IN ROBOGUIDE

- Demonstrate testing a tool frame in ROBOGUIDE.
- List the considerations for creating a user frame in ROBOGUIDE.
- List the considerations for creating a program in ROBOGUIDE.
- Demonstrate using a flash drive to transfer a program from ROBOGUIDE to a robot.

Learner Resources
atplearningresources.com
Access Code: **762316**

SECTION 10.1 CREATING A ROBOGUIDE CELL

ROBOGUIDE is FANUC's simulation software. Most major robot manufacturers have created simulation software that enables a robot programmer to create programs offline (away from the robot). There are many reasons to create programs this way.

To start, before even purchasing a robot, a programmer can create a virtual robot cell and then the program to verify that the robot can meet the process requirements. The programmer can also test several different scenarios for the fastest cycle time. All this can be done prior to purchasing any equipment. **See Figure 10-1.**

Using ROBOGUIDE Software

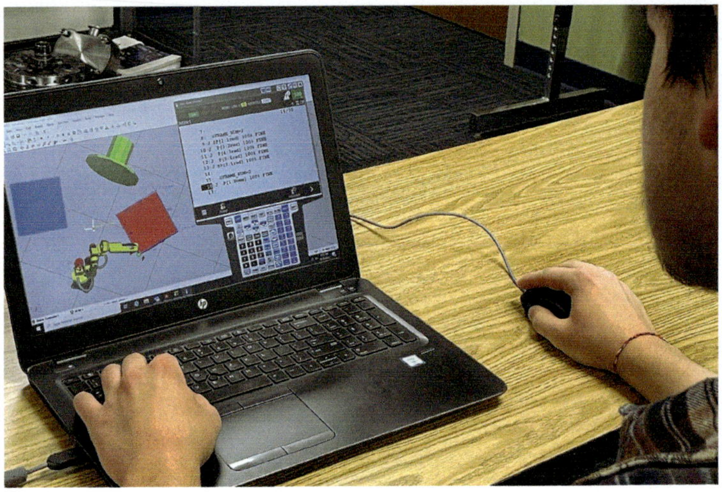

Figure 10-1. The ROBOGUIDE software allows users to practice robot movements and develop programs without access to an actual robot.

Second, a programmer can improve programs, add additional robot tasks, or add additional part numbers to the virtual robot without stopping the process that an actual robot is working on. The only time the programmer will need to stop production is to load and then touch up the new program, which takes a fraction of the time it would take to complete the work using the actual robot.

The third advantage is the speed of creating a program. The software offers a variety of shortcuts that are not available using the teach pendant. These shortcuts can drastically reduce the time to create, test, and select the best practices for desired applications.

Fourth, related to the three previously mentioned advantages, is that the software saves time and money when planning a robot cell. For example, it would be discouraging and costly to purchase a robot only to find out that the robot reach is not long enough to meet the needs of the application. Returning a robot is time-consuming and expensive as shipping and restocking fees can be thousands of dollars. Being able to try a robot cell setup virtually ensures that the right equipment can be purchased the first time.

The last advantage is in education. Inexperienced robot operators and programmers crash robots. Using the ROBOGUIDE software to train students before exposing them to an actual robot reduces the number and severity of subsequent crashes.

Virtual Robot Cells

The first thing to do in ROBOGUIDE is to create a virtual robot cell. It is best if beginners start with a simple cell that has few options. As one becomes more advanced in programming and using the ROBOGUIDE software, the cells can be more complex. For this example, a cell will contain one robot with an end effector and three fixtures (tables) selected from the CAD library within the ROBOGUIDE software.

Robot Model Selection

The robot chosen for the virtual cell should match the model that students will use in the hands-on lab. In this example, an M10 robot with integrated wiring will be chosen. If another robot needs to be chosen, then the sizes of equipment and positioning will need to be altered.

The ROBOGUIDE software is typically opened from a desktop shortcut or via the Start menu. In the Start menu, the various software modules are in the FANUC Robotics folder. The user clicks on the module, such as WeldPRO, that was licensed. When the software loads, the initial screen displays any previously created cells that can be opened. **See Figure 10-2.** In this example, a new cell will be created.

Existing Cell Selection

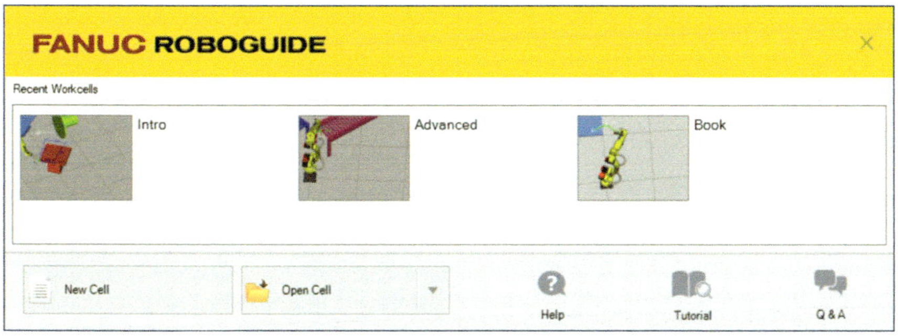

Figure 10-2. When starting the ROBOGUIDE software, users can choose to open previously saved robot cells if any are available.

After the user clicks on New Cell, the Workcell Creation Wizard appears with the process that must be completed in order to select various options for creating a robot. Then once the robot is created, more steps must be taken to add fixtures (tables), parts, and additional equipment if desired. **See Figure 10-3.**

1. Workcell Name: Name the new cell and click Next. Consider a name that is appropriately descriptive so that it can be easily identified later.
2. Robot Creation Method: Since this is the first robot cell to be created, the only option available is to create a robot using the default software configuration selected. Click Next.

   If an existing cell were available, this cell could be duplicated by selecting *Create an exact copy of an existing robot.* Choosing this option only recreates the robot and equipment. The programs contained in the robot would not be copied to the new cell.
3. Robot Software Version: This step will likely be skipped by the software because the version was already determined by the installer that was downloaded and the version that was selected during the installation process.

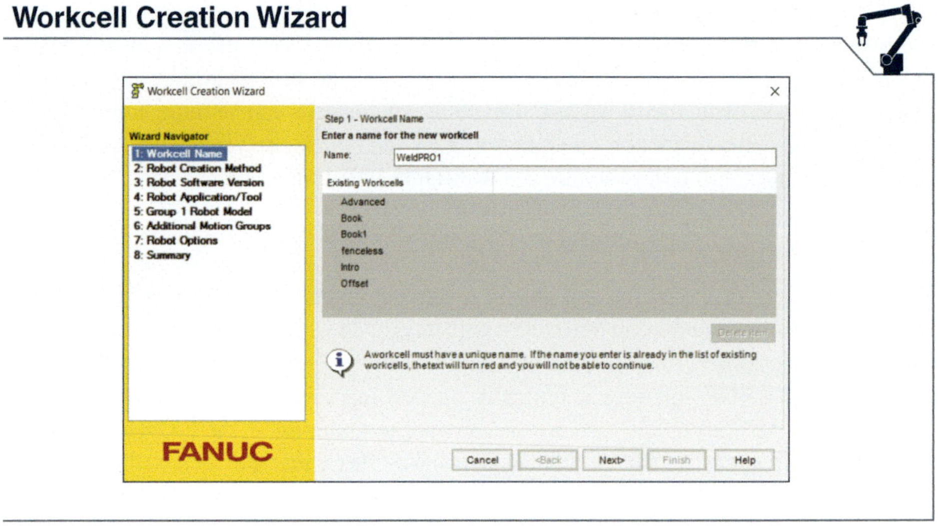

Figure 10-3. The Workcell Creation Wizard is a configuration process for creating a new robot cell.

4. Robot Application/Tool: This step involves selecting and adding an end effector. The default software has many end-effector options available. **See Figure 10-4.** The software used in this example is WeldPRO. To change to a material-handling option, select HandlingTool. Clicking the CAD Library icon shows the default end effectors available. Each item in this list can be expanded (by clicking on the + symbol) to show the end effectors under each heading. Under Grippers, there are pictures of the types available to choose from.

   Many grippers look similar, for example 36005f-200, 36005f-200-2, 36005f-200-3, and 36005f-200-4. These are all the same end effector shown in different open/close positions. The programmer can choose the different versions at various points in a program to give the illusion of the end effector opening and closing, though this added complexity is not necessary for beginners.

   For this example, choose 36005f-200 and click OK. This will apply that gripper to the robot. Then click Next.

5. Group 1 Robot Model: The specific model of robot is selected in this screen. (Group 1 refers to the servo motors that run the robot.) For the purpose of this lab, an M10 robot is selected. **See Figure 10-5.** There are several versions of M10 robots available, each with different options, such as a longer arm (to increase the robot work envelope) or through-arm cabling (which is end-effector cabling that runs through the robot arm instead of being attached outside). The one for this cell is the M10iA/12. Select this robot and click Next.

6. Additional Motion Groups: Robots can be equipped with additional groups of servo motors. Examples of additional groups might be turntables, conveyors, or rotating devices. An additional group could also be a second robot added to the cell but controlled through the same controller. An additional group must contain at least one servo motor controlled by the robot controller.

   No additional groups are required for this cell, so simply click Next to bypass this step.

7. Robot Options: All available software options for the robot selected in Step 1 are shown here. Additional software includes options such as Vision, DCS, and Constant Path. These are commonly used in industry and will often be equipped on the robot the programmer is working with. For this cell, additional software is not required. Click Next without selecting an option.

End Effector Selection

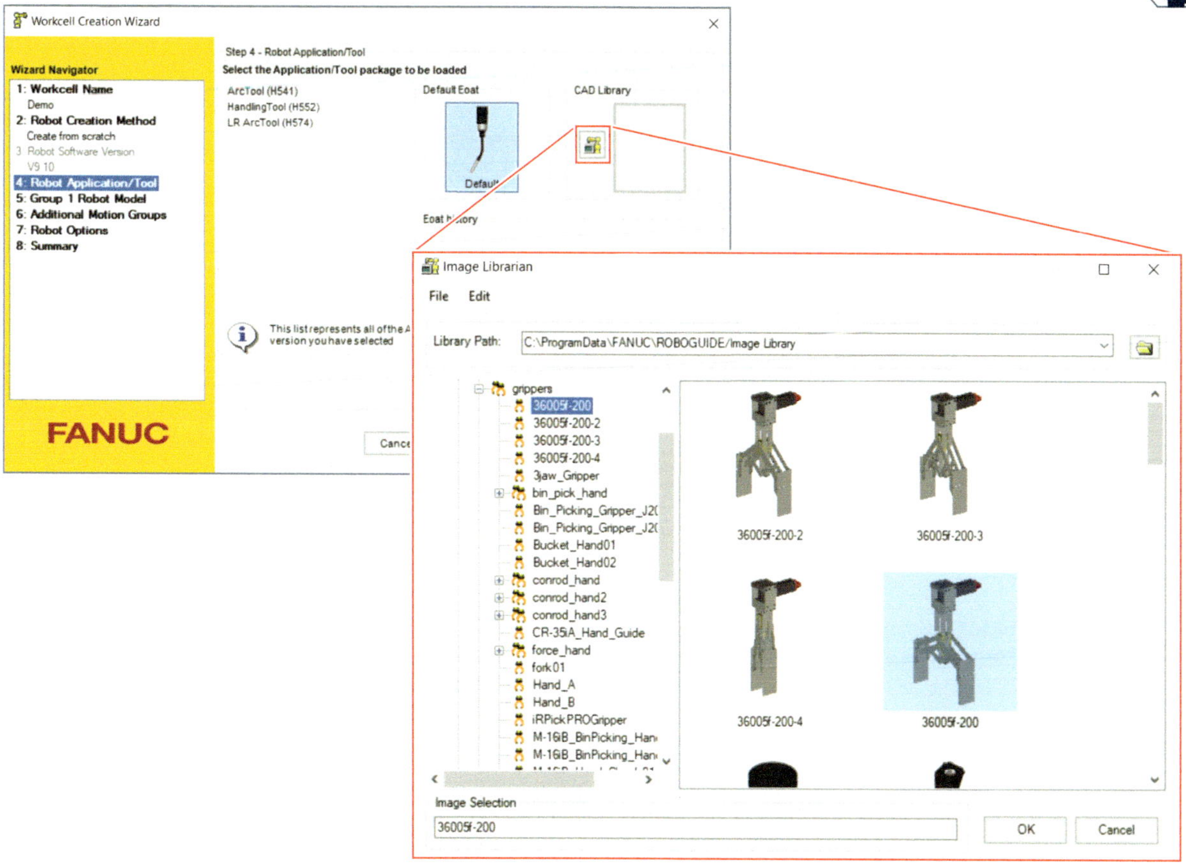

Figure 10-4. Users can select an end effector for a robot from a CAD library of existing models.

Robot Model Selection

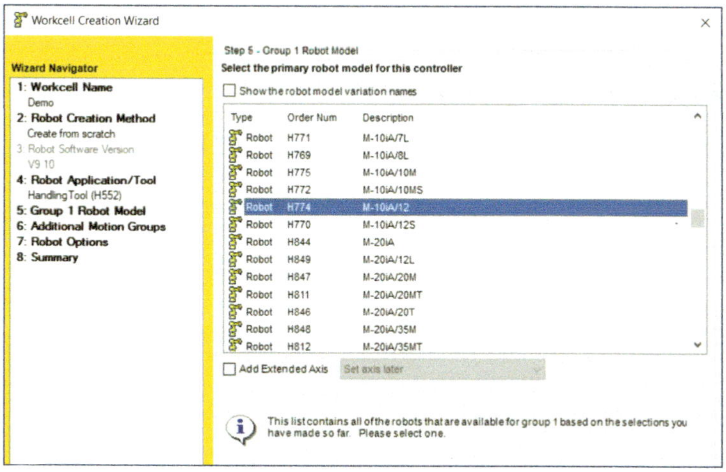

Figure 10-5. The robot for a new cell is selected from a list of available models, each of which has slightly different features.

8. Summary: This step verifies that the programmer has selected all options properly. If necessary, the programmer can use the back button to return to a previous step and change a selection. For example, one of the options that the programmer might want is Singularity Avoidance. To make this selection, return to Step 7 *Robot Options*, scroll down to *Singularity Avoidance*, and check its box. **See Figure 10-6.**

   If the option requires additional software packages not currently available, a screen will appear showing the additional software that must be added. At this point, the programmer can decide whether or not to include the software needed to eliminate the error.

When robot selections are complete, clicking Finish on the Summary screen begins the process of assembling the robot and creating the robot cell. This can take several minutes, and it is important that no keys are pressed until user input is needed.

Robot Software Options

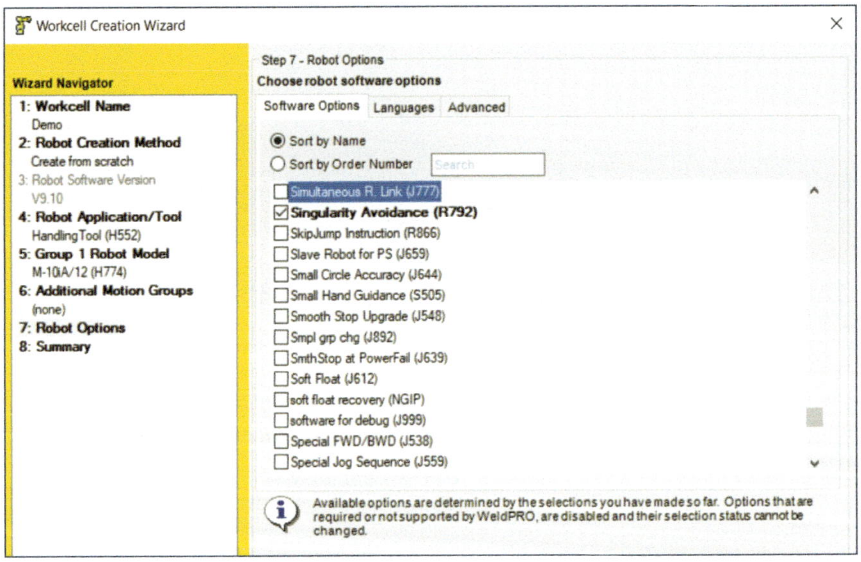

Figure 10-6. Additional software options may be loaded for individual robot cells.

Robot Initialization Options

A virtual teach pendant will appear on screen, and the display will show the initialization progress. During this process, a few additional options may require input from the programmer. The number and types of options may vary depending on the robot and software module selected.

For the example robot shown, the first option asks about the flange type, which refers to the connection pattern at the wrist. **See Figure 10-7.** The normal flange should be selected by clicking the number 1 and then clicking Enter (either on the on-screen keypad or on the physical keyboard).

Initialization Options

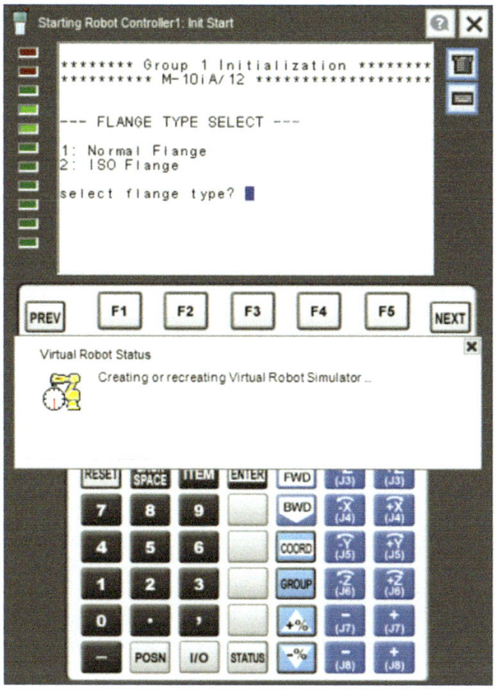

Figure 10-7. While a robot cell is initializing, it may be necessary to respond to a few additional configuration questions.

The second option asks the user to verify the type of robot selected, if using the WeldPRO module. Choice 2, M-10iA/12, should be selected for the model chosen in this example. Then click Enter.

The third option asks the user for information on the cable dress for the end effector. Number 1, integrated cable dress, should be selected for this option. Then click Enter.

The final selection asks the user to define the maximum range of motion for J1 rotation. For this option, choose 2 for *–180 . . 180[deg]* and click Enter. After this selection, give the computer several minutes to create the robot cell.

Now the configured robot and cell appears. However, the gripper is not seated properly on the robot and is nearly the size of the robot. Some settings need to be adjusted to fix this. **See the *Setting the Initial Robot Conditions* Procedure.**

PROCEDURE

Setting the Initial Robot Conditions

Lock the Robot into Position

1. Double-click on the green triad. (The triad is the three-dimensional line showing the directions of X, Y, and Z.) A properties window for the controller appears.

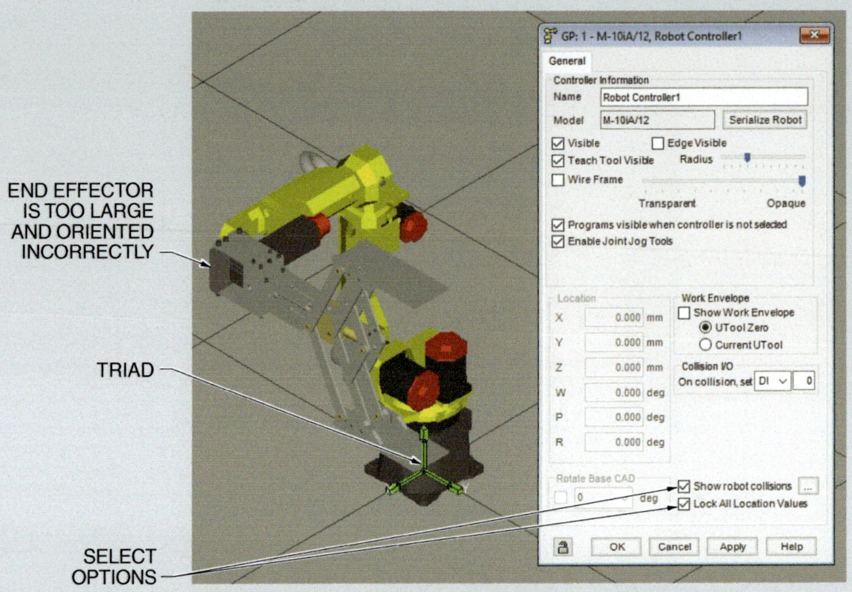

2. Select the *Show robot collisions* option. This indicates collisions by changing the robot's color.
3. Select the *Lock All Location Values* option and click OK. This option ensures that the robot's position within the cell is not inadvertently changed.

Position, Scale, and Color the End Effector

4. Double-click on the end effector. A properties window for the end effector appears, which is used to correct its position and size. There are many tabs at the top of the window for different categories of options. The current tab selected is *General*.

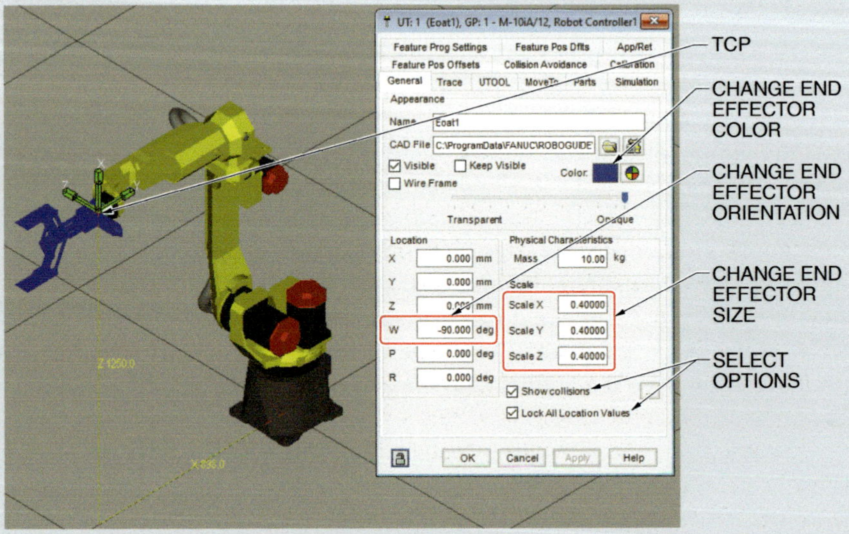

(5) In the box next to the W, type –90 and then click Apply. The end effector should change position on the robot.

(6) Under *Scale*, change the values for X, Y, and Z to 0.4 and click Apply. This changes the size of the end effector to fit the selected M10 robot. (If a smaller robot were selected, the end effector would possibly need to be made even smaller.)

(7) Click on the color wheel, select a desired color, and click OK. This is optional but makes the end effector more visible on the gray background.

(8) Select the *Lock All Location Values* option and click Apply. This locks the end effector in position on the robot so that it cannot be inadvertently moved.

Set the TCP

(9) Click on the *UTOOL* tab at the top of the properties window. From this tab, the programmer can set the robot TCP. (The TCP can also be set using direct entry from the teach pendant.)

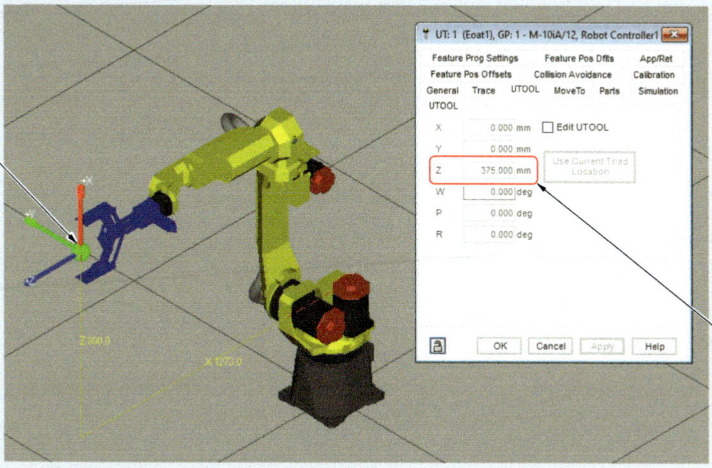

(10) Select *Edit UTOOL*.

(11) Type 375 into the Z box and click Apply. Because this end effector is centered on the robot, only the Z value needs to be changed. The green ball that represents the TCP position moves to the point on the end effector where the work is done.

(12) Click OK to close the window.

Adding Fixtures

Fixtures are added to a cell for learning how to program a robot to interact with fixed objects, such as tables. It is important to distinguish a fixture from a part in ROBOGUIDE. Fixtures can be locked in position, as was done with the robot and end effector. Parts cannot be locked and are designed in the software to be moved.

ROBOGUIDE offers a CAD library with many predesigned fixture objects that can be added to a cell, such as conveyors, pallets, tables, racks, and shelves. However, it would be impossible for this library to contain all possible options. To accommodate unique equipment utilized in different processes, custom CAD files can be loaded that represent the exact equipment in an actual robot cell. Taking the time to design and incorporate unique equipment in the robot simulation cell helps predict accurate robot or part cycle time and develop efficient robot motions. However, for simpler applications or learning environments, the CAD library of objects is sufficient. For this example cell, three tables will be added. **See the *Adding a Fixture to a Cell* Procedure.**

PROCEDURE

Adding a Fixture to a Cell

① From the *Cell* menu at the top of the ROBOGUIDE window, highlight *Add Fixture* and then click on *CAD Library*.

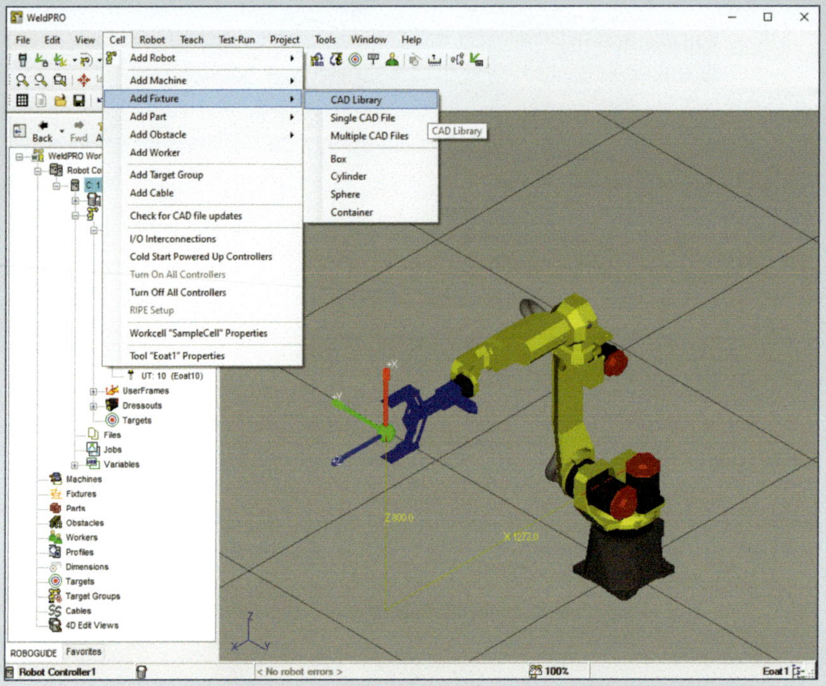

② Click on the + symbol in front of *Table* to show the selections available.

③ Click on *Container_Table*.

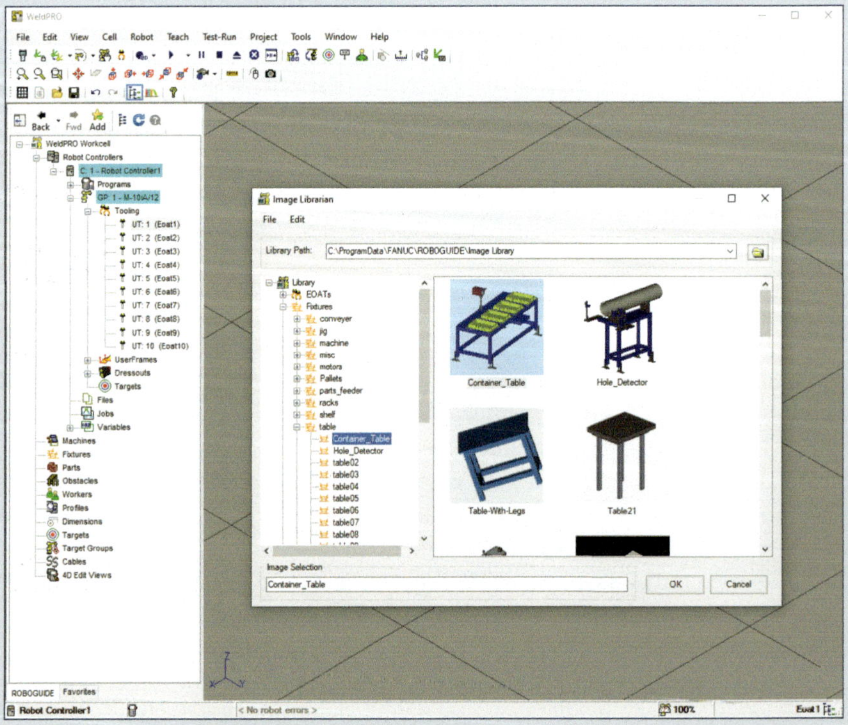

④ Scroll down to *table05*. Click on it to highlight the table and click OK.

⑤ Zoom out (typically with the center mouse wheel while the cursor is over the robot). Notice that the table is hovering high above the robot.

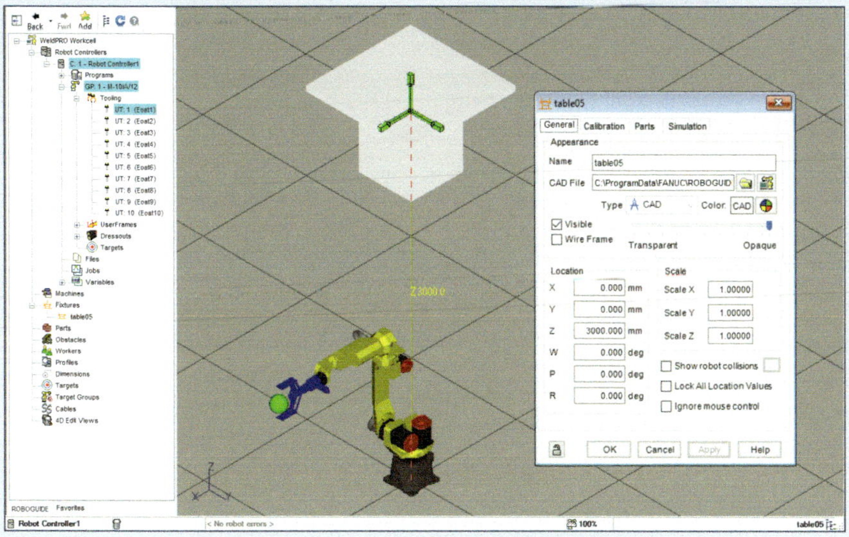

⑥ Double-click on the table to open its properties window.

⑦ Under *Location*, set the following values:

- X = 700
- Y = 500
- Z = 0
- W = 0
- P = –20
- R = –30

⑧ Under *Scale*, set the following values:

- X = 0.7
- Y = 0.7
- Z = 0.7

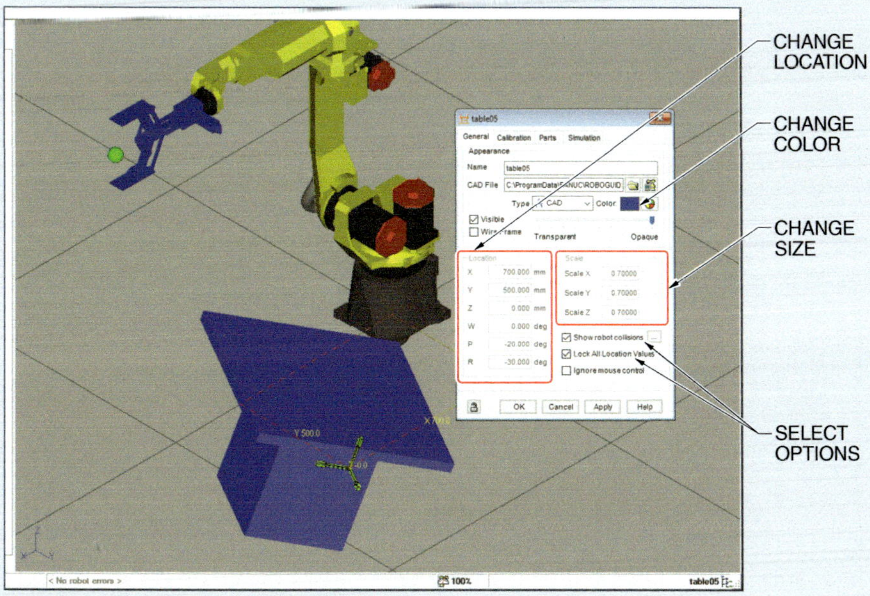

⑨ Click on the color wheel, select a blue color, and click OK.

⑩ Select *Show robot collisions* and *Lock All Location Values*.

⑪ Click Apply and OK. The table now appears to the left of the robot when seen from above.

⑫ To add a second table, repeat steps 1–11 with the following values:

- CAD Image table05
- Location X = 500
Y = –800
Z = 0
W = 0
P = 0
R = –40
- Scale X = 0.7
Y = 0.7
Z = 0.7
- Color Red
- Selected Show robot collisions
Lock All Location Values

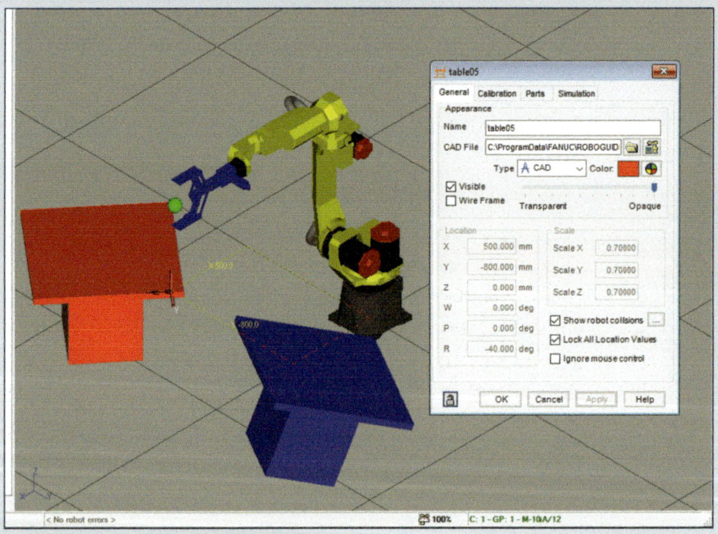

⑬ To add a third table, repeat steps 1–11 with the following values:

- CAD Image table18
- Location X = 2000
Y = 0
Z = 900
W = 0
P = –90
R = 0
- Scale X = 0.9
Y = 0.9
Z = 0.9
- Color Green
- Selected Show robot collisions
Lock All Location Values

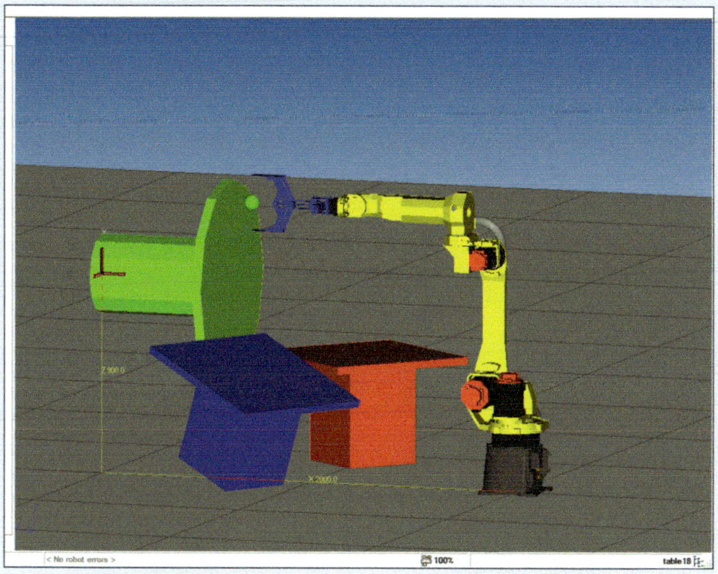

The cell is now complete, and the programmer can create user and jog frames as required and begin to program.

SECTION 10.2 NAVIGATING ROBOGUIDE

The ROBOGUIDE software has powerful capabilities that an experienced programmer can use to reduce setup and programming time by hours. It is also quite accessible for beginning programmers who take the time to learn a few basic features first. However, some command shortcuts can lead to problems if used incorrectly. While gaining experience with the software, programmers should learn the advanced shortcuts and what to watch for to ensure they will not create issues with their program.

Viewing Options

Viewing a 3D environment on a 2D monitor can be difficult for beginners. The first challenge is learning to rotate and pan the view so that the cell can be seen from all directions. This is important for ensuring the proper orientation and precise position of the robot and other objects. **See Figure 10-8.** Failure to do this could lead to the robot being out of position.

The second challenge is to ensure that robot movement is neither inefficient nor impossible. It is easy to become so focused on precise positioning that a programmer can lose track of the robot movement between positions. Problems may not be caught until the programmer tries to run the program and the robot either has a Position Not Reachable fault or it does wasteful loops and twists between points. Proficient use of the 3D viewing commands helps avoid such problems.

To be skilled with this software, the programmer should use a mouse. There are several actions that a mouse provides that would be cumbersome to accomplish otherwise.

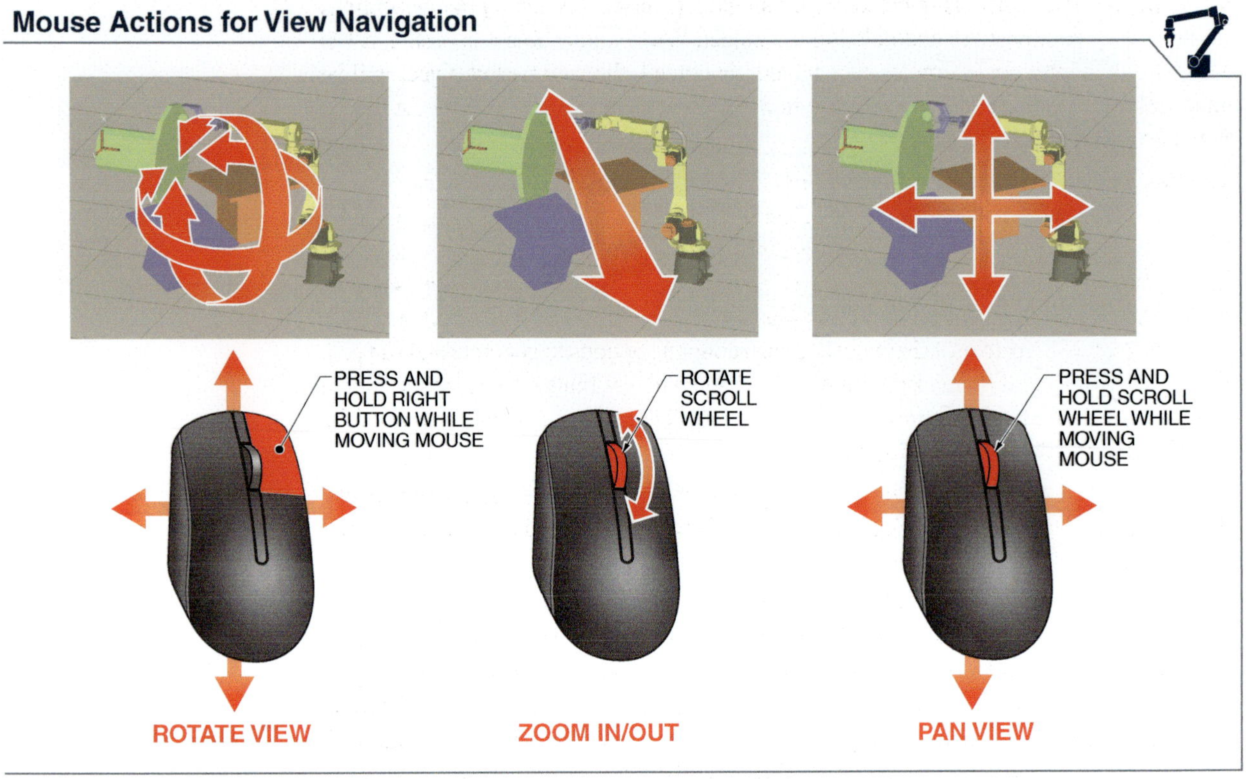

Figure 10-8. The mouse buttons are used to easily navigate the 3D environment of the robot cell.

Rotate View. With the arrow located over the robot cell, moving the mouse while holding the right mouse button down enables the programmer to rotate the view. Rotating the view is critical to seeing the robot and its TCP position from all angles.

Zoom In/Out. Rotating the scroll wheel toward the programmer zooms the robot cell out, and rotating the scroll wheel away from the programmer zooms in. Also, the zoom feature is centered at the position of the arrow. When zooming in, the programmer typically places the arrow over the item in the cell that they wish to have a closer view of before rotating the wheel.

Pan View. Pressing and holding the scroll wheel while moving the mouse will enable the programmer to pan the view (move the view without rotating it). If the programmer does not have the arrow accurately placed when zooming in, the view may shift off target. When this happens, the pan option can be used to quickly adjust the view.

Select. The left mouse button is used to make selections. Holding the arrow over the desired option and clicking the left mouse button will select that option.

Open Properties Window. Double-clicking with the left button while holding the arrow over certain objects will bring up property windows for those objects.

Menus

As expected for complicated software, the menus contain many commands and options for viewing, modifying, and testing the robot cell. However, a beginning programmer needs to know only a few of the most common ones in order to get started.

File Menu. The File menu provides the option to open other cells that the programmer may have created. This enables the programmer to open a cell without closing the ROBOGUIDE software. Opening a cell with this method is faster than closing ROBOGUIDE and then reopening it.

Edit Menu. After an object is selected (with a single left click), this menu provides options for cutting, copying, pasting, and deleting the object. When a new object is pasted, a 1 is added after its name. For example, after table05 is copied, the new pasted object will be named table051. With the Multiple Copy option, the programmer can enter the location and orientation of each fixture added.

View Menu. The View menu contains commands for changing the view of the cell and for showing additional menus or windows. Zoom and standard view commands allow the programmer to change views without using a mouse.

The Cell Browser toggles the visibility of a sidebar menu tree of all the items within the cell. The Navigator brings up a shortcut menu window for starting a new cell or adding tools, fixtures, and parts to an existing cell. It also contains options to generate and run programs. The toolbars and Status Bar commands toggle their visibility. The toolbars contain small icons as shortcuts to common menu commands. The status bar at the bottom of the ROBOGUIDE window displays any robot errors.

Cell Menu. The Cell menu provides the programmer with shortcuts for adding robots, fixtures, and parts to a robot cell. It also has controller commands and commands to open the properties windows for selected objects.

Robot Menu. The Robot menu includes the option to activate the virtual teach pendant and also provides a shortcut for viewing robot alarms and program timers. Another key command, Show Work Envelope, displays a semitransparent volume of space containing all points that the robot arm can reach, though this does not include the added reach of the end effector. **See Figure 10-9.** In this example, portions of the tables are shown outside the work envelope, but the selected end effector will allow the robot to reach all parts of the tables.

Show Work Envelope Command

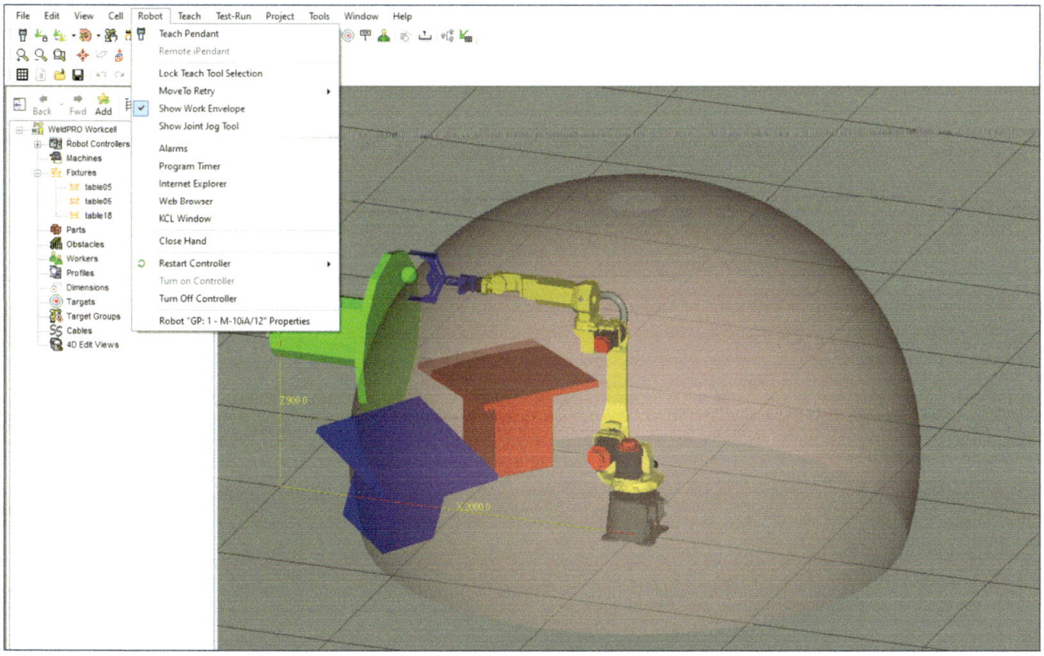

Figure 10-9. From the Robot menu, the Show Work Envelope command displays a semitransparent volume of space containing all points that the robot arm can reach.

Teach Menu. The Teach menu provides a shortcut for creating, loading, and editing TP programs. This menu also provides options for a select few editing commands, such as Find and Replace.

Test-Run Menu. The Test-Run menu provides a shortcut for running TP programs. Run options include the ability to turn on taught paths (a trace of the TCP's movements for the whole program) and the ability to loop the selected program.

Project Menu. The Project menu provides shortcuts for creating, importing, and exporting selected files. The programmer can select multiple files at once when importing or exporting the files.

Tools Menu. The Tools menu provides shortcuts for programming positions. The programmer can select positions and change their characteristics, such as movement type, speed, and termination.

Window Menu. From the Window menu, the programmer can change the number and positions of viewing panes, which enables the programmer to see the robot from multiple views simultaneously. A few different arrangements are available. **See Figure 10-10.** A multi-pane view can be useful for certain tasks, but since the panes are small, this feature is not well suited for detailed work.

Multi-Pane View

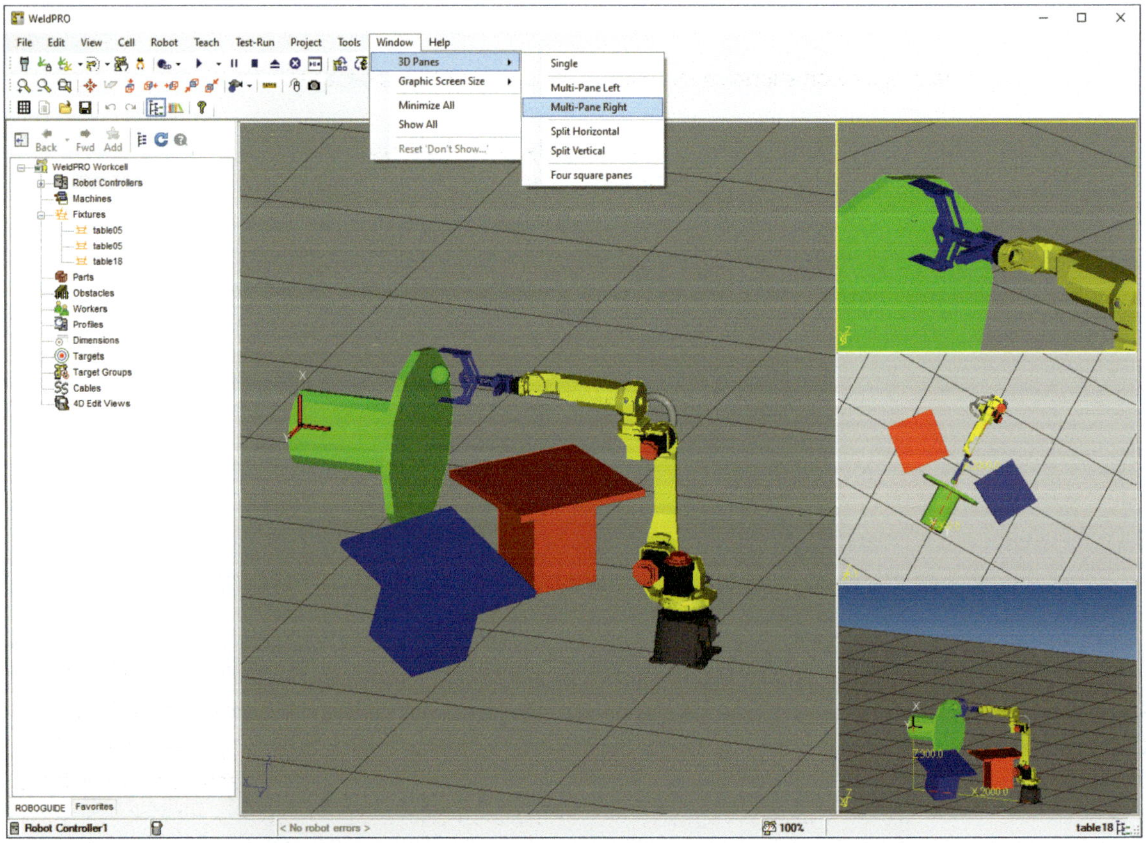

Figure 10-10. The Window menu provides more than one view of the robot cell at once.

Toolbars

Near the top of the ROBOGUIDE window are three toolbars consisting of many small icons. The icons are shortcuts for the most common options found in the menus. **See Figure 10-11.** As users become more experienced with the software, they will likely begin to use many of the toolbar shortcuts. A few deserve special mention.

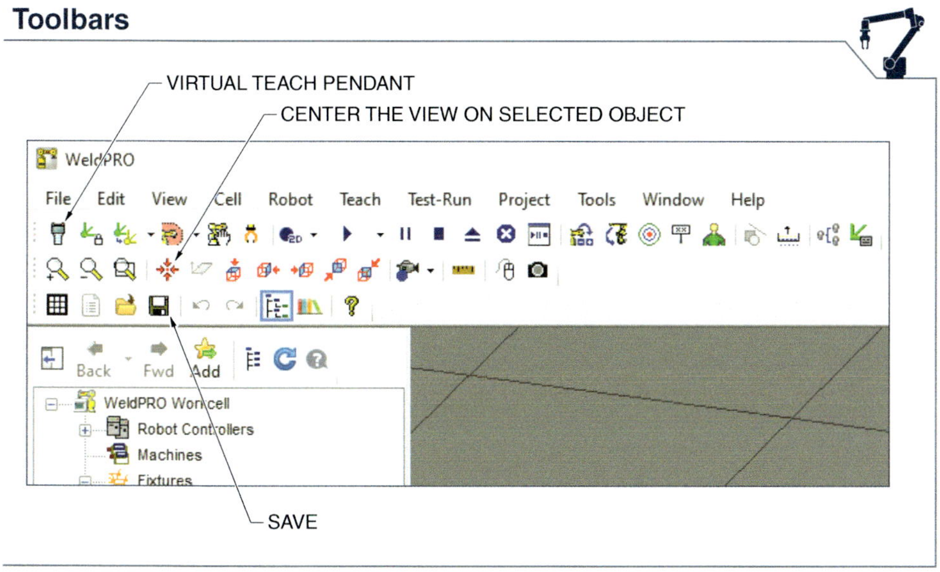

Figure 10-11. The toolbars include icon shortcuts for many menu commands and options.

The familiar Save icon enables the programmer to quickly save their work. The software does have an autosave option. However, programming can sometimes progress quickly, and to ensure no information is lost, the programmer should develop the habit of saving often.

The Center the View on Selected Object icon is particularly useful. It is common to lose track of the robot off the edge of the viewing pane while working in 3D. Clicking on this centering icon will bring the robot back to the center of the pane.

Virtual Teach Pendant

Probably the most commonly used icon shortcut is for showing/hiding the virtual teach pendant (which is also accessed via the Robot menu). This enables the user to interact with and program the virtual robot in the same manner as if they were working on an actual robot.

USB Teach Pendant

FANUC also provides the option to purchase a USB-corded teach pendant that can be used with ROBOGUIDE. This provides the student with a teach pendant exactly like what they would have on an actual robot. Because there are so many buttons and options, an experienced programmer usually relies on muscle memory to quickly jog, set up, and program a robot, which requires a lot of practice time with a teach pendant. Using a real teach pendant with the simulation software provides this practice, helping a student learn faster but without the cost and space requirements of an actual robot.

This makes the virtual teach pendant a great educational tool. Most classes have more students than robots, which limits teach pendant time for each student. As one student works on the actual robot, others can work with the software using virtual teach pendants. This way, they can all gain experience at once with the functions of the robot and teach pendant while testing their ideas. Also when working with virtual teach pendants, students cannot make a mistake that leads to a crashed robot and equipment damage.

When the teach pendant icon is clicked, the virtual teach pendant appears within the viewing pane. However, by default, it is so large that it covers the cell, and some of the teach pendant's buttons may not be visible. **See Figure 10-12.** To move the virtual teach pendant so it does not block the view and so all the buttons can be seen, click on the button that allows the teach pendant window to move outside of the main ROBOGUIDE window.

Virtual Teach Pendant

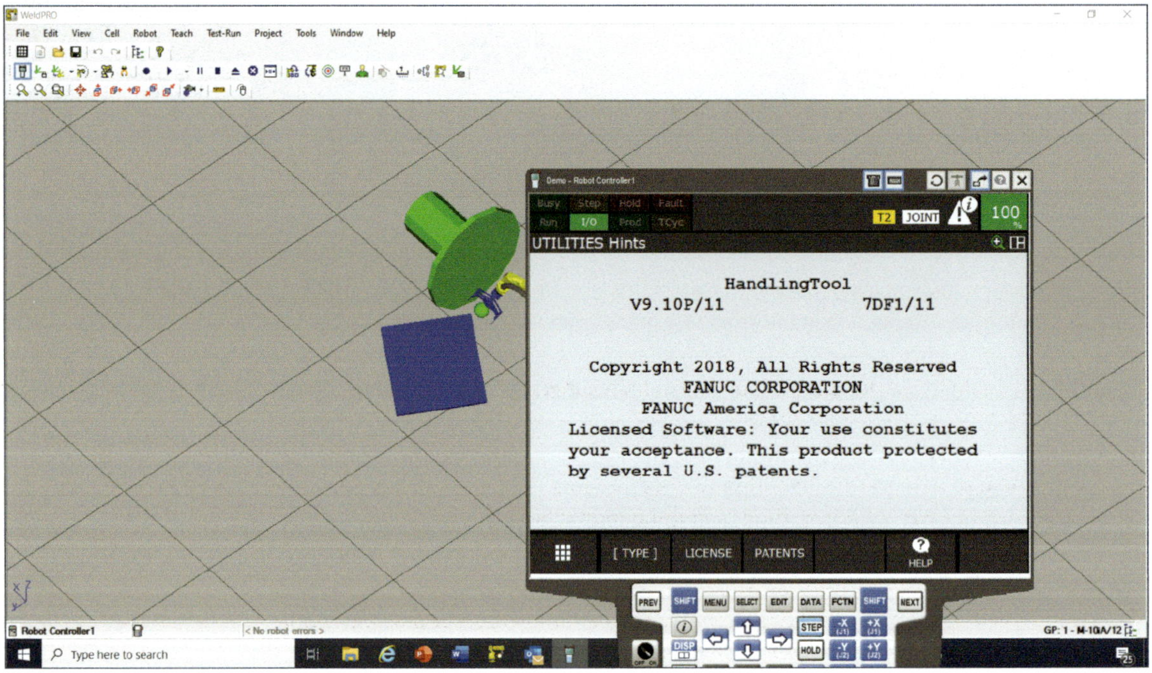

Figure 10-12. When first loaded, the virtual teach pendant may obscure the view of the robot, but the pendant can be resized and repositioned.

The display and button area can also be resized. When the cursor hovers over certain sides or corners, resize arrows appear. Left click and hold while moving the mouse to adjust the teach pendant. However, it is possible to distort the shape of the teach pendant so that some features are no longer be visible.

There are some differences between the actual teach pendant and the virtual teach pendant that should be mentioned. The first is that there are no function buttons (F-buttons) showing on the virtual keypad. **See Figure 10-13.** Function buttons are soft buttons, meaning that the purpose of each button changes as the screens change. The function buttons are still on the virtual teach pendant, but instead of showing both the keypad buttons and the screen function descriptions, only the screen function descriptions are shown, and these are clickable.

Comparing Virtual and Actual Teach Pendants

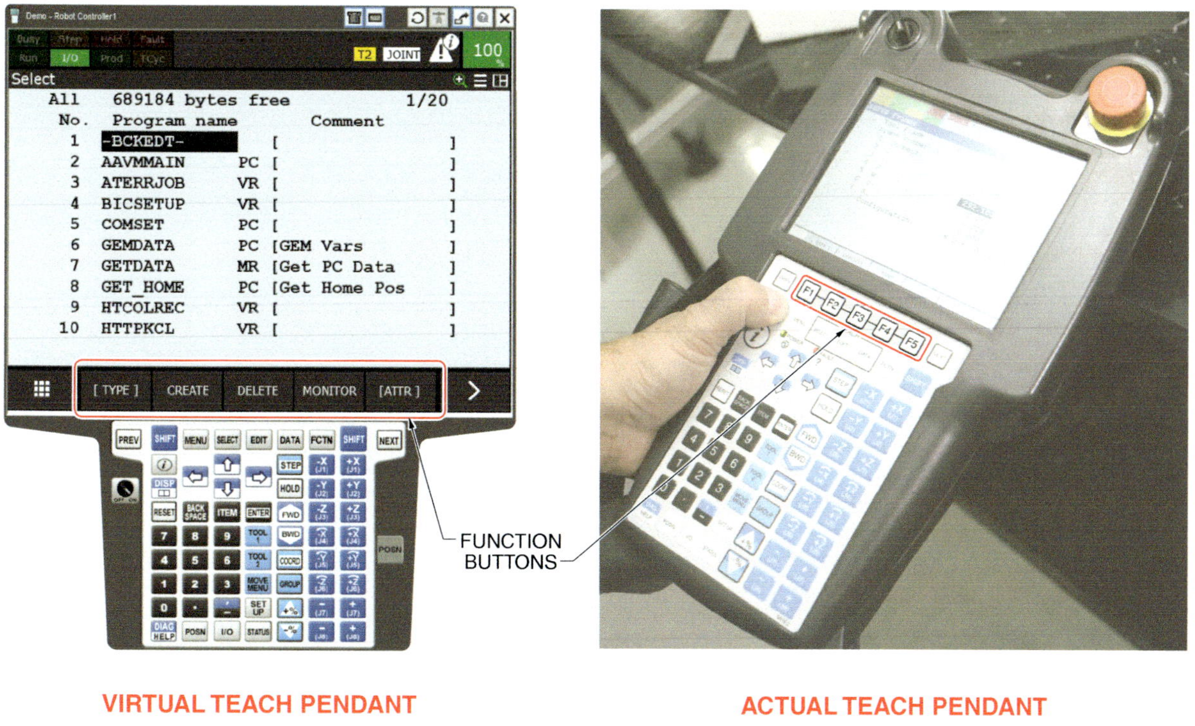

Figure 10-13. The virtual teach pendant is identical to an actual teach pendant with a few exceptions, the most obvious of which is the treatment of the function buttons.

The ON/OFF switch on the virtual pendant works like that on the actual pendant. In order to interact with the virtual robot, the teach pendant's ON/OFF switch must be in the ON position. This goes for jogging, programming, and testing a program. However, when testing a program, the virtual robot runs at actual speeds (since safety is not an issue). When testing a program with the actual teach pendant on a real robot, the robot maximum speed is 250 mm/sec or 10% of full speed.

On the virtual teach pendant, there are no deadman switches. The physical switches on the real teach pendant are a safety measure to immediately stop the robot if the programmer is distracted or injured. Because the virtual robot poses no hazards, there is no need for deadman switches.

On the actual pendant, the programmer must hold SHIFT while pressing an axis button in order to jog. If either button is released, the robot stops. This type of button action is called momentary. However, on the virtual robot, the SHIFT and axis buttons are latching. In other words, they stay "pressed" even when the mouse button is released, and they must be clicked a second time to be "unpressed." When clicked, buttons become shaded to show that they are in an activated state. As on the real robot, only one of the two SHIFT buttons must be pressed to jog.

On the real teach pendant, when the programmer is typing a comment or name, the entire teach pendant screen can be made into a keyboard. This procedure is first described in Chapter 3 when naming a tool frame. When doing this with the virtual teach pendant, a separate window appears. **See Figure 10-14.** The programmer can then use the computer keyboard to enter the desired comment or name.

Text Entry on Virtual Teach Pendants

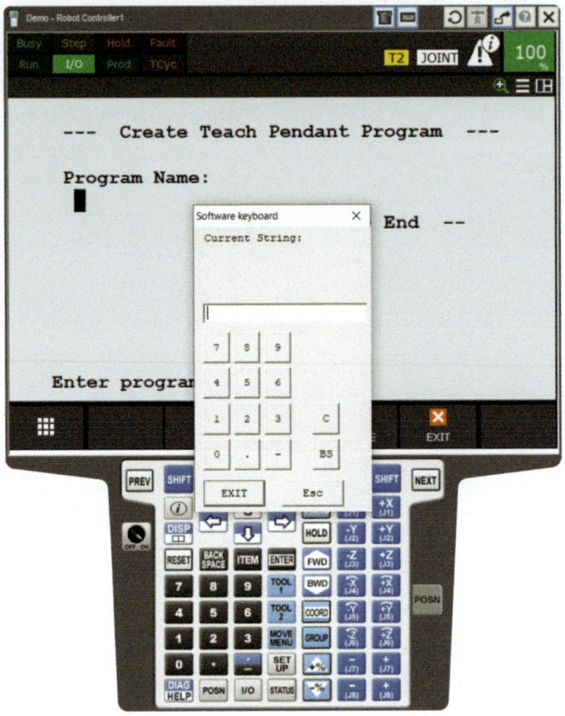

Figure 10-14. When a programmer enters text on a virtual teach pendant, a separate keypad or keyboard window may appear for easier data entry.

SECTION 10.3 PROGRAMMING IN ROBOGUIDE

Programming in ROBOGUIDE is almost exactly the same as programming with actual robots. The procedures are basically the same, with minor differences, and the virtual teach pendant is almost identical to an actual teach pendant. As with an actual robot, establishing the tool frame is the first step. Once the TCP is set, user frames are created for each fixture the robot will interact with. Then, the programmer can begin developing programs.

Testing a Tool Frame in ROBOGUIDE

The TCP is represented in ROBOGUIDE by a green ball, which helps the programmer visually verify that the TCP is properly set. With a virtual teach pendant, the process for creating a tool frame, which sets the location of the TCP, is exactly the same as when using an actual teach pendant and robot. This is described in Chapter 3. Any of the three methods discussed (direct-entry, three-point, and six-point) may be used. However, there must be a fixture in the virtual robot cell to serve as a reference for the latter two methods. A tool frame will already be created if an end effector is chosen from the CAD library, since the model includes the necessary dimensions.

On the real robot, a tool frame is tested by jogging the minor axes while in World and verifying that the end-effector rotates around an invisible point. However, testing a tool frame in ROBOGUIDE is much easier. Because the TCP is not invisible but represented by a green ball, it is easy to verify that it is in the proper position in relation to the end effector. Also, the direct-entry method can be used to adjust the position, which is readily apparent when the green ball moves. **See the *Testing a Tool Frame in ROBOGUIDE* Procedure.**

PROCEDURE

Testing a Tool Frame in ROBOGUIDE

Follow steps 1 through 12 of the procedure *Establishing a Tool Frame Using the Direct-Entry Method* in Chapter 3. After following the prompts in Section 10.2, Tool Frame 1 should show a value of 375 for the Z offset with all other offsets equal to 0. Note the position of the TCP (green ball) in relation to the end effector.

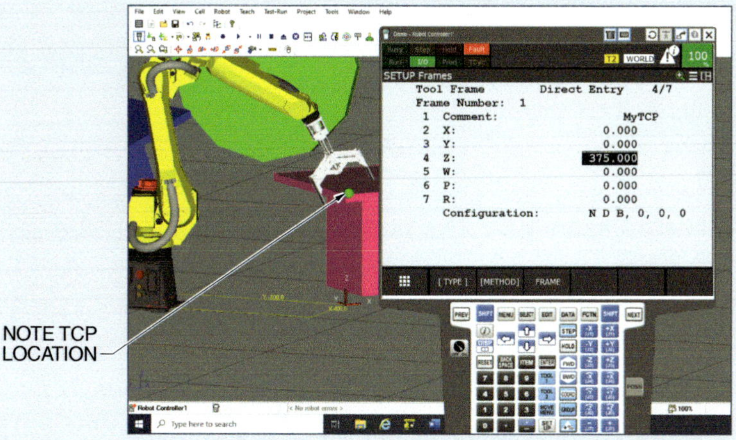

13. Use the arrow buttons to select the Z offset value and click ENTER.
14. Enter a new value for the Z offset value, such as 600, and click ENTER.

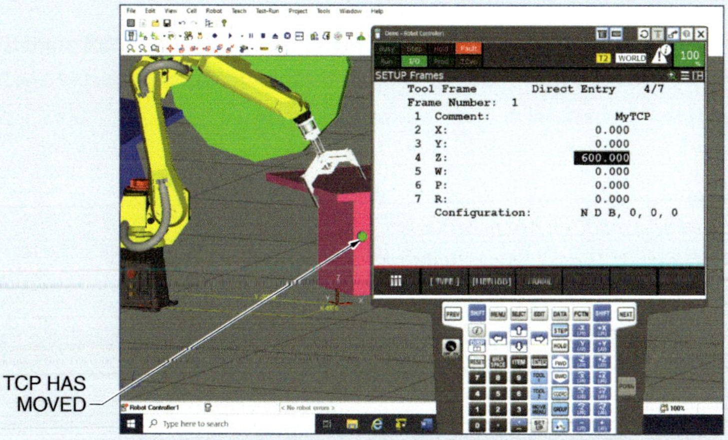

15. Note how the position of the TCP changes. As needed, use the mouse to rotate the view around the end effector.
16. Change the Z offset back to its original value (375) and note the TCP return to its proper position at the end effector.

Creating a User Frame in ROBOGUIDE

As with tool frames, creating user frames in ROBOGUIDE is done with the same procedures used on an actual robot, which are described in Chapter 3. There are several helpful tips, though, for ensuring a usable and accurate user frame, especially when using the three-point method.

- Make fixtures different colors. Then use those colors to name the associated user frames. This helps the programmer know which frame goes with which table.
- Jog the robot in World while creating the user frames.

- Use the right-hand rule to help select the origin, X, and Y points, which ensures that the Z+ direction moves the robot up and away from the fixture when jogged.
- Zoom in very close and rotate the view around the TCP to ensure that it is in the desired location. From only one view, it is easy to mistakenly believe that the TCP is in the correct position when it is actually not. **See Figure 10-15.**

Checking TCP Positions

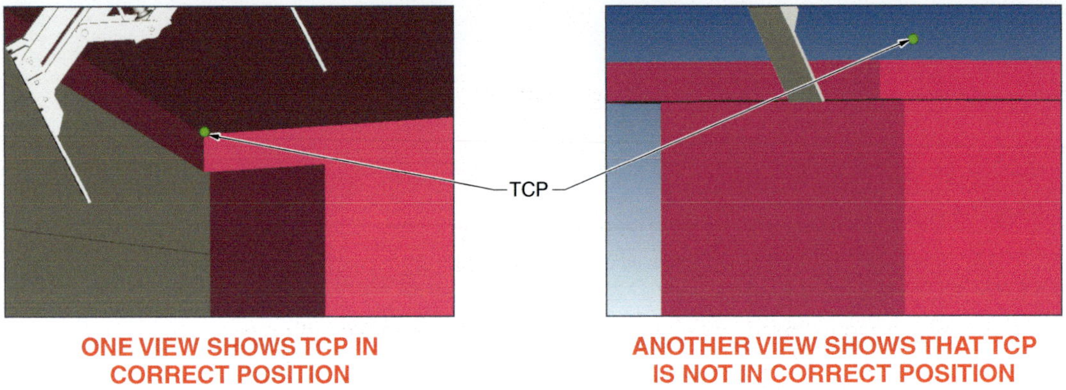

Figure 10-15. TCP positions may not always be exactly where they appear to be. Rotating the view may reveal that more adjustments are needed.

- Slow the robot down to a speed of very fine to make small adjustments at a time. Then rotate the view as needed and make other adjustments until the TCP is in the correct location.
- Zoom out often to check that the robot is not becoming twisted. If the robot appears to be twisted, click POSN and then F2 JNT to verify the positions of each joint.
- Creating a user frame with the three-point method for a round fixture in an actual robot cell is normally tricky because there is no natural right angle to use. However, round shapes are drawn as polygons in ROBOGUIDE, and the straight edges and points can be used to define the X and Y directions. **See Figure 10-16.** This is a unique advantage to ROBOGUIDE.

User Frame Points on Round Fixtures

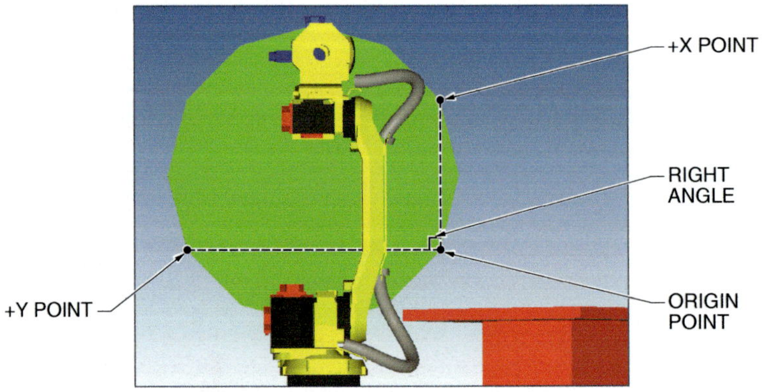

Figure 10-16. Since round objects appear as polygons in ROBOGUIDE, edge points can be used to make a precise right angle when creating a user frame with the three-point method.

Creating a Program in ROBOGUIDE

Creating a program in ROBOGUIDE is the same as creating a program on an actual robot as described in Chapter 4. While developing a program, the programmer should remember to follow a few general rules:

- Check that the robot does not get twisted into an awkward position.
- Ensure that the robot does not collide with any equipment during movements.
- Create the robot points in the proper tool and user frames.
- Select proper movement type, speed, and termination for each desired movement.

Transferring a Program to a Robot

A programmer will eventually need to transfer a program developed and tested in ROBOGUIDE to an actual robot. The simplest way to do this is to load the program onto a flash drive. The robot controller has a USB port available for copying the program from the flash drive onto the robot's internal storage.

The program file that is saved does not contain the tool frame or user frame, so those still must be created with the actual robot. Also, points will need to be touched up due to slight differences between the virtual cell and the actual cell. However, all the logic, comments, and remarks are transferred, which still saves an enormous amount of time.

To save a program created in ROBOGUIDE to a flash drive, the programmer needs a flash drive of 8 GB or smaller. While the cell to be saved is open, cell browser may already be open. (If it is not open, click *Cell Browser* from the *View* menu.) The cell browser shows all the elements and characteristics of the cell, including a list of programs. From the list of programs, right click on the desired program to save. From the menu that appears, highlight *Save* and left click. The two options are *Binary (.TP)* and *Text (.LS)*. **See Figure 10-17.** The *Text (.LS)* option saves a readable text copy of the program that can be viewed with a text editor, such as Word®. The *Binary (.TP)* option saves the program in a binary format that can be read by the robot controller but is not readable text. Select the *Binary (.TP)* option and save the file to the attached flash drive.

After the robot is turned on and allowed to boot completely, the teach pendant is used to copy the program onto the controller. **See the *Loading a TP File from a Flash Drive to a Robot* Procedure.**

Saving Program Files

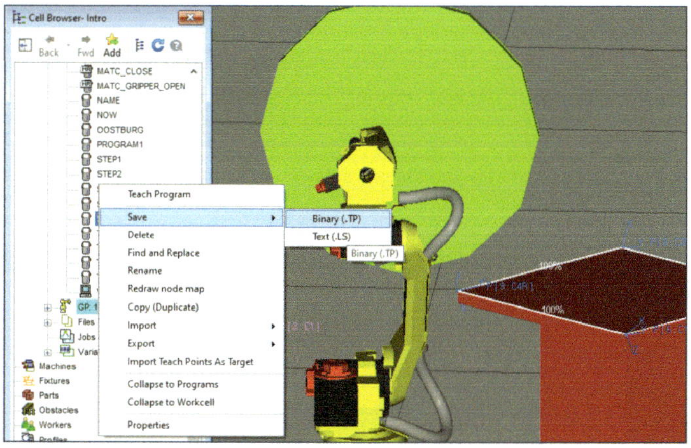

Figure 10-17. When saving a program file to a flash drive, the choice is to save it in a binary format (for loading onto a robot) or in a text format (for saving a readable copy).

PROCEDURE

Loading a TP File from a Flash Drive to a Robot

1. Insert the flash drive into the USB port on the robot controller or on the teach pendant.
2. Press MENU.
3. Highlight `FILE` and press ENTER. Note the top left of the File screen shows `MC`, which stands for "main controller."

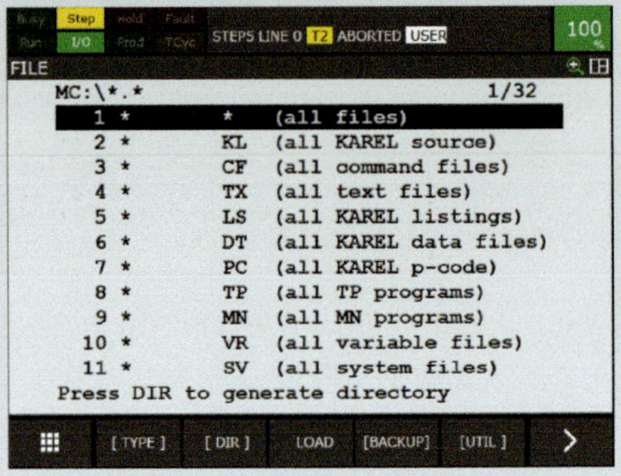

4. Press F5 `[UTIL]`.
5. Highlight `Set Device` and press ENTER.
6. From the list, select `USB DISK (UD1)` if the flash drive was inserted into the controller or `USB DISK (UT1)` if it was inserted into the teach pendant. Press ENTER.
7. Highlight `TP (all TP programs)` and press ENTER.

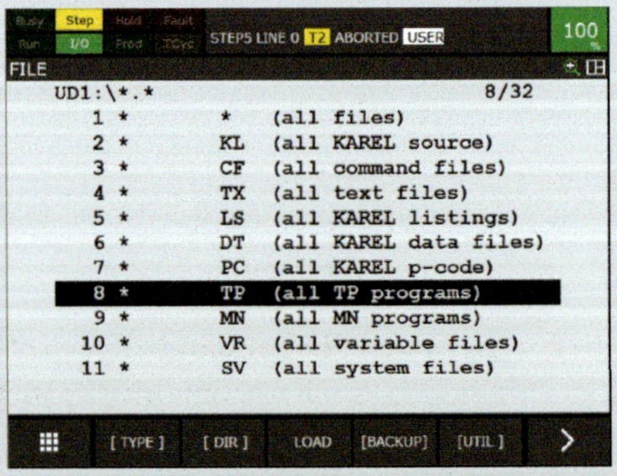

8. Press F3 `LOAD`. This copies any TP files from the flash drive to the robot controller.
9. Press SELECT. Newly loaded programs will appear in the program list.

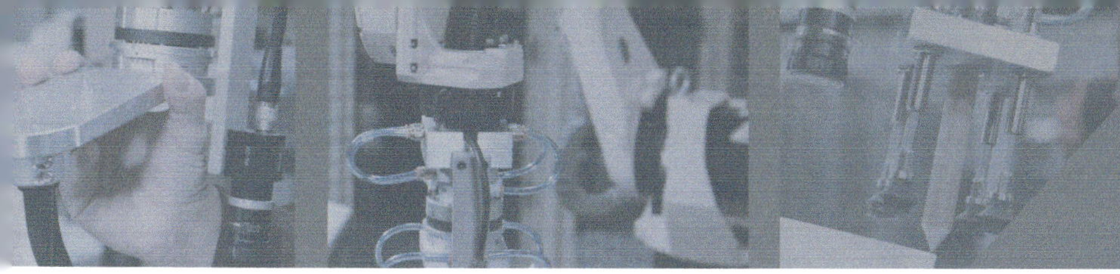

Review 10

Name ______________________________ Date ________________

SECTION 10.1 CREATING A ROBOGUIDE CELL

________ **1.** With ROBOGUIDE, a programmer can test several different scenarios for the fastest ___.

________ **2.** ROBOGUIDE offers a variety of shortcuts that are not available using the ___.

________ **3.** The first thing to do in ROBOGUIDE is to create a ___.

A. fixture
B. virtual robot cell
C. program
D. user frame

________ **4.** A programmer can choose the different versions of a(n) ___ model at various points in a program to give the illusion of it opening and closing.

________ **5.** Examples of additional ___ might be turntables, conveyors, or rotating devices.

A. cells
B. parts
C. groups
D. cycles

________ **6.** The number and types of additional robot initialization options may vary depending on the ___ and software module selected.

A. robot
B. group
C. teach pendant
D. end effector

________ **7.** ROBOGUIDE offers a(n) ___ with many predesigned fixture objects that can be added to a cell, such as conveyors, pallets, tables, racks, and shelves.

8. How does using ROBOGUIDE save time and money when planning a robot cell?

9. How is ROBOGUIDE an advantage for education?

10. What is not duplicated when creating an exact copy of an existing robot?

11. How can adding custom equipment improve the accuracy of virtual robot cell operation?

SECTION 10.2 NAVIGATING ROBOGUIDE

_____________ **12.** The programmer should use a ___ for view navigation.

A. mouse
B. keyboard
C. teach pendant
D. touch screen

_____________ **13.** Moving the mouse while holding the ___ button down enables the programmer to rotate the view.

_____________ **14.** The zoom feature is centered at the position of the ___.

_____________ **15.** The ___ is used to make selections.

A. left mouse button
B. right mouse button
C. enter key
D. scroll wheel button

_____________ **16.** The ___ toggles the visibility of a sidebar menu tree of all the items within the cell.

A. Status Bar
B. Navigator
C. Cell Browser
D. Cell Menu

_____________ **17.** The ___ menu provides a shortcut for creating, loading, and editing TP programs.

____________________ **18.** On the virtual teach pendant, the SHIFT and axis buttons are ___, meaning they stay "pressed" even when the mouse button is released, and they must be clicked a second time to be "unpressed."

19. What are the challenges of viewing a 3D environment on a 2D monitor?

20. How does the programmer open property windows for objects?

21. How is the Center the View on Selected Object command useful?

22. How are function buttons different on the virtual teach pendant?

23. Why does the virtual teach pendant have no deadman switches?

SECTION 10.3 PROGRAMMING IN ROBOGUIDE

_______________ **24.** As with an actual robot, establishing the ___ frame is the first step.

A. user
B. fixture
C. tool
D. jog

_______________ **25.** The ___ is represented in ROBOGUIDE by a green ball.

_______________ **26.** A tool frame will already be created if an end effector is chosen from the ___, since the model includes the necessary dimensions.

_______________ **27.** The simplest way to transfer a program developed and tested in ROBOGUIDE to an actual robot is to load the program onto a(n) ___.

_______________ **28.** To save a program created in ROBOGUIDE to a flash drive, the programmer needs a flash drive of ___ or smaller.

A. 8 MB
B. 32 MB
C. 8 GB
D. 32 GB

_______________ **29.** The Binary (___) option saves the program in a binary format that can be read by a robot controller but is not readable text.

A. .TP
B. .TCP
C. .BIN
D. .ROB

30. Why is testing a tool frame in ROBOGUIDE easier than with a real robot?

31. What is a trick for using the three-point method in ROBOGUIDE to create a user frame on a round shape?

32. What is not included in a program file saved from ROBOGUIDE?

10 Lab Exercise

Lab Objective

Set up virtual robot cells and create programs in ROBOGUIDE.

Equipment Required

- Installed ROBOGUIDE software

Lab Procedure

1. Create a virtual robot cell as described in this chapter. Include the red table, blue table, and green (round) table as specified.
2. Test the TCP for the selected end effector. Create user frames for each of the three tables.
3. Create a program with the following sequence:
   - Start from the Home position. (Use a position register.)
   - Move to a corner on the red table and trace its edge in a clockwise movement. All movements on this table should be linear, should have fine terminations, and should be at 45% speed.
   - Return to the Home position.
   - Move to a corner on the blue table and trace its edge in a counter clockwise movement. These movements should be joint with continuous 10 terminations.
   - Call a second program that traces the circular table immediately after tracing the blue table. The circular table trace should use circular movement with continuous 30 terminations.
4. After verifying proper program operation, add the necessary instructions to modify the program for the following changes:
   - Trace the red table four times. (Use registers and jump labels.)
   - Move immediately to the blue table and trace this table once. Then trace the round table once. Complete this pair of traces four times. (Use registers and jump labels.)
   - Complete the entire program from start to finish twice.

Note: Ensure that points on each table use the correct user frame. Only use two registers to complete Step 4.

Glossary

A

analog I/O: The set of analog inputs and outputs between a controller and other cell equipment.

analog signal: An electrical signal that varies in voltage or current in accordance with some variable parameter.

automation technician: A robot technician who programs and configures robots and may also help troubleshoot problems in robot operation.

axis: A linear or rotational direction of movement.

axis limit: A rotational position beyond which a joint cannot rotate, either by programmed or physical limits.

C

Cartesian coordinate system: A system of describing points in three-dimensional space as distances along three perpendicular axes from an origin point.

Cartesian jogging: A jogging method in which all axes operate simultaneously (as needed) in order to move the tool center point (TCP) in a straight line or to move the robot about the TCP.

circle arc movement: A movement in which the TCP moves in an arc-shaped path.

circular movement: A type of movement in which the TCP moves in a full circle.

cold start: The standard method of initializing a robot for typical programming or operations by turning on the power while not pressing any teach pendant buttons.

collaborative robot (cobot): A robot that is specifically designed to operate side-by-side with personnel.

combination end effector: An end effector composed of multiple types of tooling or grippers.

comment: A string of text added to a program to describe or explain code in a way that is easily readable.

complementary outputs: Two outputs that must always have opposite statuses (ON or OFF).

continuous termination (CNT): An instruction to a robot to allow a certain amount of deviation from a defined point in order to move smoothly through it to the next point.

controlled start: The startup procedure for powering up a robot in preparation for adding or modifying robot software and overriding errors.

controller: The power and control unit for a robot. Also known as a control cabinet.

cycle time: The time it takes a robot to complete one set of programmed actions, which is then repeated.

D

deadman switch: A switch that must be manually depressed in order for a desired operation to continue, and releasing the switch immediately stops the operation.

delta robot: A robot with a robot base and three articulated arms suspended from the base in a triangular pattern.

digital I/O: The set of digital inputs and outputs between a controller and other cell equipment.

digital signal: An electrical signal that indicates only ON or OFF (or HIGH or LOW) states.

E

emergency stop (E-stop): A condition in which a machine's controls immediately stop all movement.

end effector: A tool attached to the end of a robot arm for completing a specific task. Also known as end-of-arm-tooling (EOAT).

F

fine termination: An instruction to a robot to move to a programmed position, stop, and then accelerate to the next position.

fixture: A support device for holding items or assemblies securely in a specific position.

frame: A set of three perpendicular axes (X, Y, and Z), which intersect at an origin (zero) point.

G

group: A set of axes, such as a robot's axes, and also possibly other electromechanical devices that could be controlled by the robot program.

group I/O: A set of digital inputs and outputs that are assigned to be activated (turned ON and OFF) as a group.

group mask: A list of possible axis groups with indications of which are active and available.

H

Home position: A neutral position for a robot that minimizes the stress on the servomotors and gears.

I

industrial robot: An automated, programmable, mechanical actuator capable of performing tasks repeatedly with a high degree of precision.

input: An information signal that is supplied to a controller from other equipment.

integrator: A company that designs, installs, and programs complete robot systems for industrial applications.

J

jog frame: A robot frame with a user-defined origin and X, Y, and Z-axis directions and is used to orient the robot through the space of a robot cell.

jogging: The manual adjustment of a robot's position and orientation via the teach pendant.

joint jogging: A jogging method in which each robot joint is operated independently (with the exception of Axis 2).

joint movement: A type of movement in which the TCP is moved with the most efficient joint rotations.

jump label: A set of two instructions that directs the execution of code from one part in a program to another part of the same program.

K

KAREL program: A computer file containing lines of code written in KAREL.

KAREL: A Pascal-based programming language designed specifically for controlling FANUC robots.

L

limit switch: An ON/OFF switch that is activated when something pushes on its lever arm.

linear movement: A type of movement in which the TCP moves in a straight line.

M

macro (MR) program: A computer file containing a small section of code for a particular, commonly repeated action.

maintenance technician: A technician who maintains, repairs, and troubleshoots a robot's mechanical and electrical systems.

major axis: One of the first three axes (1, 2, or 3), which provide the largest amount of robot movement.

message: A note that appears (typically on the teach pendant) for the operator when the robot reaches a certain point in its program.

minor axis: One of the last three axes (4, 5, or 6), which provide fine robot movement.

movement line: A line of code that specifies a particular position in space and the motion specifications to be used when moving to that point.

movement type: A designation for the shape of the definable motions of a TCP.

O

offset programming: The process of programming where position registers (PRs) are modified to become the desired robot positions.

output: A control signal supplied from a controller to other equipment.

P

part-present device: An electrical switch on an end effector that senses the presence of a picked part and indicates if it is properly held.

payload: The combined mass of an end effector and any object that is picked up and handled by the end effector.

payload schedule: A list of payload masses and their associated information.

position ID: A unique number identifying a particular point in space, along with orientation angles and the associated tool frame number and user frame number.

position register (PR): A position definition saved in memory that can be accessed by any program on a robot.

preprogramming: The process of creating a robot program entirely within cell simulation software and without access to the physical robot.

pressure switch: An ON/OFF switch that senses a change in pressure through the use of a movable diaphragm.

program adjust: A procedure for making small adjustments to robot movements while it is running a program.

program adjust schedule: A list of possible position, orientation, and speed offsets to be applied to one or more consecutive movement lines of program code.

programmable logic controller (PLC): A computer system used for controlling industrial processes.

program shell: A program file containing the program's name and empty lines but does not yet contain any code.

proximity switch: An ON/OFF switch that senses the presence of a nearby object without physical contact.

R

register: A software variable used to store a number.

Remark command: A command that designates existing lines that are not to be executed by the software.

Remark instruction: An added instruction line within program code that has no effect on the robot and is intended as a plain-language note by a programmer to explain a section of code.

repeatability: The ability of a robot to return to the same positions many times within a very narrow tolerance.

right-hand rule: A hand arrangement that uses fingers as a memory aid to show the directions of the X, Y, and Z axes.

rigid attachment: An attachment of an end effector to a robot that requires significant downtime to remove.

robot arm: A robot composed of multiple rotating joints connected by rigid segments. Also known as an articulated robot.

robot cell: An all-inclusive system consisting of one or more robots and their controllers within an enclosed work area and includes any safety devices and other equipment that the robot(s) interact with.

robot I/O: The set of digital inputs and outputs between a controller and the robot end effector.

robot operator: A worker who readies a robot to perform programmed tasks, monitors its operation, and does any manual chores required when the robot completes a production run.

S

selective compliance articulated robot arm (SCARA) robot: A robot with articulated arms that are limited to movement in two dimensions.

servomotor: An electric motor with an integrated position encoder that is used to control the precise degree of rotation of an axis.

singularity: A specific robot position in which any subsequent movement can be accomplished in multiple ways and the robot controller cannot choose which one to execute.

stripped-down robot: A robot that has not yet been equipped with an end effector.

subprogram: A program file with a section of code intended to be called by other programs for a specific task.

system variable (SV) file: A computer file containing a list of parameters and their values, which can be adjusted to change the way the teach pendant and robot function.

T

teach pendant: A handheld device that interfaces with controller software and is used by the robot programmer to jog, configure, or program a robot.

teach pendant (TP) program: A computer file containing lines of code for controlling robot actions.

termination type: A specification of how accurately the robot passes through a position as it continues to the next position.

timer: A programming instruction that measures the robot's physical execution time of a section of code.

tool center point (TCP): A point centered on an end effector and defines where the robot's work is done.

tool changer: An electronically controlled module installed between a robot and an end effector that enables the robot to automatically release one end effector and attach another.

tool frame: A robot frame with an origin at the tool center point (TCP) and axes that extend from that point in three directions (X, Y, and Z).

U

user frame: A robot frame with a user-defined origin and X, Y, and Z-axis directions, which is used to orient a robot to a particular object within a robot cell, such as a fixture.

user screen: The display used for showing operator messages.

W

witness mark: A physical mark on one side of a joint that is used with degree markings on the other side of the joint to identify the zero position of that axis.

work envelope: The volume of space that is defined by the maximum distance the robot can reach.

World frame: The default Cartesian frame in which the origin is centered on and just above the rotation point of Axis 1, the X axis is pointing out from the robot's front, the Y axis is pointing to the robot's left, and the Z axis is pointing up.

Index

Page numbers in italic refer to figures.

G

H

I

J

K

L

M

O

P

R

S

T

U

V

W